Computing Skills for Biologists

Computing Skills for Biologists

....

A TOOLBOX

Stefano Allesina & Madlen Wilmes

PRINCETON UNIVERSITY PRESS
PRINCETON AND OXFORD

Copyright © 2019 by Princeton University Press

Published by Princeton University Press
41 William Street, Princeton, New Jersey 08540
6 Oxford Street, Woodstock, Oxfordshire OX20 1TR

press.princeton.edu

All Rights Reserved

Library of Congress Control Number: 2018935166
ISBN 9780691167299
ISBN (pbk.) 9780691182759

British Library Cataloging-in-Publication Data is available

Editorial: Alison Kalett, Lauren Bucca, and Kristin Zodrow
Production Editorial: Mark Bellis
Text and Cover Design: Lorraine Doneker
Cover Credit: Desmond Paul Henry, 1963, courtesy of the D. P. Henry Archive
Production: Erin Suydam
Publicity: Alyssa Sanford
Copyeditor: Alison Durham

This book has been composed in MinionPro

Printed on acid-free paper. ∞

Printed in the United States of America

1 3 5 7 9 10 8 6 4 2

*To all biologists
who think they can't code.*

• • • • • • • • • • •

Science is what we understand well enough to explain to a computer. Art is everything else we do.   —Donald E. Knuth

Summary of Contents

Contents

· · · · · · · · ·

Figures

· · · · · · ·

Acknowledgments
● ● ● ● ● ● ● ● ● ● ● ● ● ● ● ● ●

This book grew out of the lecture notes for the graduate class Introduction to Scientific Computing for Biologists, taught by Stefano at the University of Chicago. We would like to thank all the students who took the class—especially those enrolled in Winter 2016 and 2017, who have test-driven the book. The class was also taught in abbreviated form at various locations: thanks to the TOPICOS students at the University of Puerto Rico Río Piedras, to those attending the 2014 Spring School of Physics at the Abdus Salam International Center for Theoretical Physics in Trieste, to the participants in the Mini-Course BIOS 248 at the Hopkins Marine Station of Stanford University, and to the students who joined the University of Chicago BSD QBio Boot Camps held at the Marine Biological Laboratory in Woods Hole, MA.

Several people read incomplete drafts of the book, or particular chapters. Thanks to Michael Afaro and anonymous referees for the critical and constructive feedback. Alison Kalett and her team at Princeton University Press provided invaluable support throughout all the stages of this long journey.

We are grateful to all the scientists who uploaded their data to the Dryad Digital Repository, allowing other researchers to use it without restrictions: it's thanks to them that all the exercises in this book make use of real biological data, coming from real papers.

The development of this book was supported by the National Science Foundation CAREER award #1148867.

Stefano: I started programming late in my college years, thanks to a class taught by Gianfranco Rossi. This class was a real revelation: I found out that I loved programming, and that it came very naturally to me. After learning a lot from my cousin and friend Davide Lugli, I even started working as a software developer for a small telephone company. In the final year of college, programming turned out to be very useful for my honors thesis. I worked with Alessandro Zaccagnini, who introduced me to the beauty of LaTeX. When I started graduate school, my advisor Antonio Bodini encouraged me to keep working on my computing skills. Stefano Leonardi convinced me to switch to Linux, pressed me to learn C, and introduced me to R. Many people are responsible for my computational toolbox, but I want to mention Daniel Stouffer, who gave me a crash course in svn, and Ed Baskerville, who championed the use of Git. I thank my students and post-docs (in order of appearance Anna Eklöf, Si Tang, Phillip Staniczenko, Liz Sander, Matt Michalska-Smith, Samraat Pawar, Gyuri Barabás, Jacopo Grilli, Madlen Wilmes, Carlos Marcelo-Sérvan, Dan Maynard, and Zach Miller) and the other members of the lab for coping with my computational quirks and demands, and for learning with me many of the tools covered in this book. Finally, I want to thank my parents, Gianni and Grazia for buying me a computer instead of a motorcycle, my brother Giulio, and my family, Elena, Luca & Marta, for love and support.

Madlen: I earned a five-year degree in biology without a single hour of computational training. That turned out to be a tremendous problem. Fellow PhD student Illka Kronholm took the time to teach R to the "German without humor." Later, Ben Brachi generously shared his scripts and contributed to my fluency in R. I am also grateful to Marco Mambelli who introduced me to cluster computing and helped me to get a grip on Unix.

My advisor and coauthor Stefano Allesina had, undoubtedly, the biggest impact on my programming skills. His course, Introduction to Scientific Computing, was my first experience of a well-structured and constructive class centered on computing skills. And so the idea for this book was born, as I wished every student could have a resource to help overcome the initial steep learning curve of many computing skills, and use examples that were actually relevant to a biologist's daily work. I am tremendously grateful to Stefano for agreeing to write this book together. In the process I not only became a more proficient programmer and better organized scientist, but also felt inspired by his productivity and positive attitude.

My dad-in-law, George Wilmes, provided valuable feedback on every chapter and took care of my kids so I could work on this book. Last but not least I want to thank my parents and my husband John for helpful suggestions, love, and support.

CHAPTER 0

• • • • • • • • • • •

Introduction: Building a Computing Toolbox

No matter how much time you spend in the field or at the bench, most of your research is done when sitting in front of a computer. Yet, the typical curriculum of a biology PhD does not include much training on how to use these machines. It is assumed that students will figure things out by themselves, unless they join a laboratory devoted to computational biology—in which case they will likely be trained by other members of the group in the laboratory's (often idiosyncratic) selection of software tools. But for the vast majority of academic biologists, these skills are learned the hard way—through painful trial and error, or during long sessions sitting with the one student in the program who is "good with computers."

This state of affairs is at odds with the enormous growth in the size and complexity of data sets, as well as the level of sophistication of the statistical and mathematical analysis that goes into a modern scientific publication in biology. If, once upon a time, coming up with an original idea and collecting great data meant having most of the project ready, today the data and ideas are but the beginning of a long process, culminating in publication.

The goal of this book is to build a basic computational toolbox for biologists, useful both for those doing laboratory and field work, and for those with a computational focus. We explore a variety of tools and show how they can be integrated to construct complex pipelines for automating data collection, storage, analysis, visualization, and the preparation of manuscripts ready for submission.

These tools are quite disparate and can be thought of as LEGO® bricks, that can be combined in new and creative ways. Once you have added a new tool to your toolbox, the potential for new research is greatly expanded. Not only will you be able to complete your tasks in a more organized, efficient, and reproducible way, but you will attempt answering new questions that would have been impossible to tackle otherwise.

0.1 The Philosophy

Fundamentally, this book is a manifesto for a certain approach to computing in biology. Here are the main points we want to emphasize:

Automation

Doing science involves repeating the same tasks several times. For example, you might need to repeat an analysis when new data are added, or if the same analysis needs to be carried out on separate data sets, or again if the reviewers ask you to change this or that part of the analysis to make sure that the results are robust.

In all of these cases you would like to automate the processing of the data, such that the data organization and analysis and the production of figures and statistical results can be repeated without any effort. Throughout the book, we keep automation at the center of our approach.

Reproducibility

Science should be reproducible, and much discussion and attention goes into carefully documenting empirical experiments so that they can be repeated. In theory, reproducing statistical analysis or simulations should be much easier, provided that the data and parameters are available. Yet, this is rarely the case—especially when the processing of the data involves clicking one's way through a graphical interface without documenting all the steps. In order to make it easy to reproduce your results, your computational work should be

readable: Your analysis should be easy to read and understand. This involves writing good code and documenting what you are doing. The best way to proceed is to think of your favorite reader: yourself, six months from now. When you receive feedback from the reviewers, and you have to modify the analysis, will you be able to understand precisely what you did, how, and why? Note that there is no way to email yourself in the past to ask for clarifications.

organized: Keeping the project tidy and well organized is a struggle, but you don't want to open your project directory only to find that there are 16 versions of the same program, all with slight—and undocumented—variations!

self-contained: Ideally, you want all of your data, code, and results in the same place, without dependencies on other files or code that are not in the same location. In this way, it is easy to share your work with others, or to work on your projects from different computers.

Openness

Science is a worldwide endeavor. If you use costly, proprietary software, the chances are that researchers in less fortunate situations cannot reproduce your results or use your methods to analyze their data. Throughout the book, we focus on *free software*:[1] not only is the software free in the sense that it costs nothing, but free also means that you have the *freedom* to run, copy, distribute, study, change, and improve the software.

Simplicity

Try to keep your analysis as simple as possible. Sometimes, "readable" and "clever" are at odds, meaning that a single line of code processing data in 14 different ways at once might be genius, but seldom is it going to be readable. In such cases, we tend to side with readability and simplicity—even if this means writing three additional lines of code. We also advocate the use of plain text whenever possible, as text is portable to all computer architectures and will be readable decades from now.

Correctness

Your analysis should be correct. This means that programming in science is very different from programming in other areas. For example, *bugs* (errors in the code) are something the software industry has learned to manage and live with—if your application unexpectedly closes or if your word processor sometimes goes awry, it is surely annoying, but unless you are selling pacemakers this is not going to be a threat. In science, it is essential that your code does solely what it is meant to do: otherwise your results might be unjustified. This strong emphasis on correctness is peculiar to science, and therefore you

---

1. gnu.org/philosophy/free-sw.html.

will not find all of the material we present in a typical programming textbook. We explore basic techniques meant to ensure that your code is correct and we encourage you to rewrite the same analysis in (very) different programming languages, forcing you to solve the problem in different ways; if all programs yield exactly the same results, then they are probably correct.

Science as Software Development

There is a striking parallel between the process of developing software and that of producing science. In fact, we believe that basic tools adopted by software developers (such as version control) can naturally be adapted to the world of research. We want to build software pipelines that turn ideas and data into published work; the development of such a pipeline has important milestones, which parallel those of software development: one can think of a manuscript as a "beta version" of a paper, and even treat the comments of the reviewers as bugs in the project which we need to fix before releasing our product! The development of these pipelines is another central piece of our approach.

0.2 The Structure of the Book

The book is composed of 10 semi-independent chapters:

Chapter 1: Unix
> We introduce the Unix command line and show how it can be used to automate repetitive tasks and "massage" your data prior to analysis.

Chapter 2: Version control
> Version control is a way to keep your scientific projects tidily organized, collaborate on science, and have the whole history of each project at your fingertips. We introduce this topic using Git.

Chapter 3: Basic programming
> We start programming, using Python as an example. We cover the basics: from assignments and data structures to the reading and writing of files.

Chapter 4: Writing good code
> When we write code for science, it has to be correct. We show how to organize your code in an effective way, and introduce debugging, unit testing, and profiling, again using Python.

Chapter 5: Regular expressions

> When working with text, we often need to find snippets of text matching a certain "pattern." Regular expressions allow you to describe to a computer what you are looking for. We show how to use the Python module re to extract information from text.

Chapter 6: Scientific computing

> Modern programming languages offer specific libraries and packages for performing statistics, simulations, and implementing mathematical models. We briefly cover these tools using Python. In addition, we introduce Biopython, which facilitates programming for molecular biology.

Chapter 7: Scientific typesetting

> We introduce LaTeX for scientific typesetting of manuscripts, theses, and books.

Chapter 8: Statistical computing

> We introduce the statistical software R, which is fully programmable and for which thousands of packages written by scientists for scientists are available.

Chapter 9: Data wrangling and visualization

> We introduce the tidyverse, a set of R packages that allow you to write pipelines for the organization and analysis of large data sets. We also show how to produce beautiful figures using ggplot2.

Chapter 10: Relational Databases

> We present relational databases and sqlite3 for storing and working efficiently with large amounts of data.

Clearly, there is no way to teach these computational tools in 10 brief chapters. In fact, in your library you will find several thick books devoted to each and every one of the tools we are going to explore. Similarly, becoming a proficient programmer cannot be accomplished by reading a few pages, but rather it requires hundreds of hours of practice. So why try to cover so much material instead of concentrating on a few basic tools?

The idea is to provide a structured guide to help jump-start your learning process for each of these tools. This means that we emphasize breadth over depth (a very unusual thing to do in academia!) and that success strongly depends on your willingness to practice by trying your hand at the exercises and embedding these tools in your daily work. Our goal is to *showcase* each tool by first explaining what the tool is and why you should master it. This allows you to make an informed decision on whether to invest your time in learning how to use it. We then guide you through some basic features and give you a step-by-step explanation of several simple examples. Once you have worked through these examples, the learning curve will appear less steep, allowing you to find your own path toward mastering the material.

0.2.1 How to Read the Book

We have written the book such that it can be read in the traditional way: start from the first page and work your way toward the end. However, we have striven to provide a modular structure, so that you can decide to skip some chapters, focus on only a few, or use the book as a quick reference.

In particular, the chapters on Unix (ch. 1), version control (ch. 2), LaTeX (ch. 7), and databases (ch. 10) can be read quite independently: you will sometimes find references to other chapters, but in practice there are no prerequisites. Also, you can decide to skip any of these chapters (though we love each of these tools!) without affecting the reading of the rest of the book.

We present programming in Python (chs. 3–6) and then again in R (chs. 8–9). While we go into more detail when explaining basic concepts in Python, you should be able to understand all of the R material without having read any of the other chapters. Similarly, if you do not plan to use R, you can skip these chapters without impacting the rest of the book.

0.2.2 Exercises and Further Reading

In each chapter, upon completion of the material you will be ready to start working on the "Exercises" section. One of the main features of this book is that exercises are based on real biological data taken from published papers. As such, these are not silly little exercises, but rather examples of the challenges you will overcome when doing research. We have seen that some students find this level of difficulty frustrating. It is entirely normal, however, to have no idea how to solve a problem at first. Whenever you feel that frustration is blocking your creativity and efficiency, take a short break. When you return, try breaking the problem into smaller steps, or start from a blank slate and attempt an entirely different approach. If you keep chipping away at the exercise, then little by little you will make sense of what the problem entails and—finally—you will find a way to crack it. Learning how to enjoy problem solving and to take pride in a job well done are some of the main characteristics of a good scientist.

For example, we are fond of this quote from Andrew Wiles (who proved the famous Fermat's last theorem, which baffled mathematicians for centuries): "You enter the first room of the mansion and it's completely dark. You stumble around bumping into the furniture but gradually you learn where each piece of furniture is. Finally, after six months or so, you find the light switch, you turn it on, and suddenly it's all illuminated."[2] Hopefully, it will take you less than six months to crack the exercises!

Note that there isn't "a" way to solve a problem, but rather a multitude of (roughly) equivalent ways to do so. Each and every approach is perfect, provided that the results are correct and that the solution is found in a reasonable amount of time. Thus, we encourage you to consult our solutions to the exercises only once you have solved them: Did we come up with the same idea? What are the advantages and disadvantages of these approaches? Even if you did not solve the task entirely, you have likely learned a lot more while trying, compared to reading through the solutions upon hitting the first stumbling block. To provide a further stepping stone between having no idea where to start and a complete solution, we provide a pseudocode solution of each exercise online: the individual steps of the solution are described in English, but no code is provided. This will give you an idea how to approach the problem, but you will need to come up with the code. From there, it is only a short way to tackling your very own research questions. You can find the complete solutions and the pseudocode at computingskillsforbiologists.com/exercises.

When solving the exercises, the internet is your friend. Finding help online is by no means considered "cheating." On the contrary, if you find yourself exploring additional resources, you are doing exactly the right thing! As with research, anything goes, as long as you can solve your problem (and give credit where credit is due). Consulting the many comprehensive online forums gives you a sense of how widespread these computational tools are. Keep in mind that the people finding clever answers to your questions also started from a blank slate at some point in their career. Moreover, seeing that somebody else asked exactly your question should further convince you that you are on the right track.

Last but not least, the "Reading" section of each chapter contains references to books, tutorials, and online resources to further the knowledge of the material. If the chapter is an appetizer, meant to whet your appetite for knowledge, the actual meal is contained in the reading list. This book is a road map that equips you with sufficient knowledge to choose the appropriate tool for each task, and take the guesswork out of "Where should I start

---

2. computingskillsforbiologists.com/provingfermat.

my learning journey?" However, only by reading more on the topic and by introducing these tools into your daily research work will you be able to truly master these skills, and make the most of your computer.

We conclude with the sales pitch we use to present the class that inspired this book. If you are a graduate student and you read the material, you work your way through all the exercises, constantly striving to further your knowledge of these topics by introducing them into your daily work, then you will shave six months off your PhD—and not *any* six months, but rather those spent wrestling with the data, repeating tedious tasks, and trying to convince the computer to be reasonable and spit out your thesis. All things considered, this book aims to make you a happier, more productive, and more creative scientist. Happy computing!

0.3 Use in the Classroom

We have been teaching the material covered in this book, in the graduate class Introduction to Scientific Computing for Biologists, at the University of Chicago since 2012. The enrollment has been about 30 students per year. We found the material appropriate for junior graduate students as well as senior undergraduates with some research experience.

The University of Chicago runs on a quarter system, allowing for 10 lectures of three hours each. Typically, each chapter is covered by a single lecture, with "Version Control" (ch. 2) and "Scientific Typesetting" (ch. 7) each taking about an hour and a half, and "Writing Good Code" (ch. 4) and "Statistical Computing" (ch. 8) taking more than one lecture each.

In all cases, we taught students who had computers available in class, either by teaching in a computer lab, or by asking students to bring their personal laptops. Rather than using slides, the instructor lectured while typing all the code contained in the book during the class. This makes for a very interactive class, in which all students type all of the code too—making sure that they understand what they are doing. Clearly, this also means that the pace is slowed down every time a student has included a typo in their commands, or cannot access their programs. To ease this problem, having teaching assistants for the class helps immensely. Students can raise their hand, or stick a red post-it on their computer to signal a problem. The teaching assistant can immediately help the student and interrupt the class in case the problem is shared by multiple students—signaling the need for a more general explanation.

To allow the class to run smoothly, each student should prepare their computer in advance. We typically circulate each chapter a week in advance of class, encouraging the students to (a) install the software needed for the class and (b) read the material beforehand. Teaching assistants also offer weekly office hours to help with the installation of software, or to discuss the material and the exercises in small groups.

The "intermezzos" that are interspersed in each chapter function very well as small in-class exercises, allowing the students to solidify their knowledge, as well as highlighting potential problems with their understanding of the material.

We encourage the students to work in groups on the exercises at the end of each chapter, and review the solutions at the beginning of the following class. While this can cause some difficulties in grading, we believe that working in groups is essential to overcome the challenge of the exercises, making the students more productive, and allowing less experienced students to learn from their peers. Publishing a blog where each group posts their solutions reinforces the esprit de corps, creating a healthy competition between the groups, and further instilling in the students a sense of pride for a job well done. We also encouraged students to constructively comment on the different approaches of other groups and discuss the challenges they've faced while solving the exercises.

Another characteristic of our class has been the emphasis on the practical value of the material. For example, we ask each student to produce a final project in which they take a boring, time-consuming task in their laboratory (e.g., analysis of batches of data produced by laboratory machines, calibration of methods, other repetitive computational tasks) and completely automate it. The student then shows their work to their labmates and scientific advisor, and writes a short description of the program, along with the documentation necessary to use it. The goal of the final project is simply to show the student that mastering this material can save them a lot of time—even when accounting for the strenuous process of writing their first programs.

We have also experimented with a "flipped classroom" setting, with mixed results. In this case, the students read the material at their own pace, and work through all the small exercises contained in the chapter. The lecture is then devoted to working on the exercises at the end of each chapter. The lecturer guides the discussion on the strategies that can be employed to solve the problem, sketching pseudocode on the board, and eventually producing a fully fledged code on the computer. We have observed that, while this approach is very rewarding for students with some prior experience in programming, it is much less engaging for novices, who feel lost and out of touch with the rest of the class. Probably, this would work much better if the class size were small (less than 10 students).

Finally, we have found that leading by example serves as powerful motivation to students. We have always shown that we use the tools covered here for our own research. A well-placed anecdote on Git saving the day, or showing how all the tables in a paper were automatically generated with a few lines of R, can go a long way toward convincing the students that their work studying the material will pay off over a lifetime.

0.4 Formatting of the Book

You will find all commands and the names of packages typeset in a fixed-width font. User-provided [INPUT] is capitalized and set between square brackets. To execute the commands, you do not need to reproduce such formatting. Within explanatory text, *technical terms* are presented in italics.

Throughout the book, we provide many code examples, enclosed in gray boxes and typeset using fixed-width fonts. All code examples are also provided on the companion website computingskillsforbiologists.com—but we encourage you to type all the code in by yourself: while this might feel slow and inefficient, the learning effect is stronger compared to simply copying and pasting, and only inspecting the result. Within the code examples, language-specific commands are highlighted in bold.

Within the code boxes, we try to keep lines short. When we cannot avoid a line that is longer than the width of the page we use the symbol ↳ to indicate that what follows should be typed in the same line as the rest.

0.5 Setup

Before you can start computing, you need to set up the environment, and download the data and the code.

What You Need

A computer: All the software we present here is free and can be installed with a few commands in Linux Ubuntu or Apple's OS X; we strive to provide guidance for Windows users. There are no specific hardware requirements. All the tools require relatively little memory and space on your hard drive.

Software: Each chapter requires installing specific software. We have collected detailed instructions guiding you through the installation of each tool at computingskillsforbiologists.com/setup.

A text editor: While working through the chapters, you will write a lot of code. Much will be written in the integrated development environments (IDEs) Jupyter and RStudio. Sometimes, however, you will need to write code in a text editor. We encourage you to keep working with your favorite editor, if you already have one. If not, please choose an editor that can support syntax highlighting for Python, R, and LATEX. There are many options to choose from, depending on your architecture and needs.[3]

Initial Setup

You can find instructions for the initial setup on our website at computing skillsforbiologists.com/setup. We have bundled all the data, code, exercises, and solutions in a single download. We strongly recommend that you save this directory in your home directory (see section 1.3.2).

---

3. computingskillsforbiologists.com/texteditors.

CHAPTER 1

• • • • • • • • • • • •

Unix

1.1 What Is Unix?

Unix is an operating system, which means that it is the software that lets you interface with the computer. It was developed in the 1970s by a group of programmers at the AT&T Bell laboratories. The new operating system was an immediate success in academic circles, with many scientists writing new programs to extend its features. This mix of commercial and academic interest led to the many variants of Unix available today (e.g., OpenBSD, Sun Solaris, Apple's OS X), collectively denoted as *nix systems. Linux is the open source Unix clone whose "engine" (kernel) was written from scratch by Linus Torvalds with the assistance of a loosely knit team of hackers from across the internet. Ubuntu is a popular Linux distribution (version of the operating system).

All *nix systems are multiuser, network-oriented, and store data as plain text files that can be exchanged between interconnected computer systems. Another characteristic is the use of a strictly hierarchical file system, discussed in section 1.3.2.

This chapter focuses primarily on the use of the Unix shell. The shell is the interface that is used to communicate with the core of the operating system (kernel). It processes the commands you type, translates them for the kernel, and shows you the results of your operations. The shell is often run within an application called the *terminal*. Together, the shell and terminal are also referred to as a *command-line interface* (CLI), an interface that allows you to input commands as successive lines of text. Though technically not correct, the terms shell, command line (interface), and terminal are often used interchangeably. Even if you have never worked with a command-line interface, you have surely seen one in a movie: Hollywood likes the stereotype of a hacker typing code in a small window with a black background (i.e., command-line interface).

Today, several shells are available, and here we concentrate on the most popular one, the Bash shell, which is the default shell in Ubuntu and OS X. When working on the material presented in this book, it is convenient, though not strictly necessary, to work in a *nix environment. Git Bash for Windows emulates a Unix shell. As the name Git Bash implies, it also uses the Bash shell.

1.2 Why Use Unix and the Shell?

Many biologists are not familiar with using *nix systems and the shell, but rather prefer graphical user interfaces (GUIs). In a GUI, you work by interacting with graphical elements, such as buttons and windows, rather than typing commands, as in a command-line interface. While there are many advantages to working with GUIs, working in your terminal will allow you to automate much of your work, scale up your analysis by performing the same tasks on batches of files, and seamlessly integrate different programs into a well-structured pipeline.

This chapter is meant to motivate you to get familiar with command-line interfaces, and to ease the initially steep learning curve. By working through this chapter, you will add a tool to your toolbox that is the foundation of many others—you might be surprised to find out how natural it will become to turn to your terminal in the future. Here are some more reasons why learning to use the shell is well worth your effort:

First, Unix is an operating system written by programmers for programmers. This means that it is an ideal environment for developing your code and managing your data.

Second, hundreds of small programs are available to perform simple tasks. These small programs can be strung together efficiently so that a single line of Unix commands can perform complex operations, which otherwise would require writing a long and complicated program. The ability to create these pipelines for data analysis is especially important for biologists, as modern research groups produce large and complex data sets whose analysis requires a level of automation and reproducibility that would be hard to achieve otherwise. For instance, imagine working with millions of files by having to open each one of them manually to perform an identical task, or try opening your single 80 Gb whole-genome sequencing file in software with a GUI! In Unix, you can string a number of small programs together, each performing a simple task, and create a complex pipeline that can be stored in a script (a text file containing all the commands). Such a script makes your work 100% reproducible. Would you be able to repeat the exact series of 100 clicks of a complex analysis in a GUI? With a script, you will always obtain the same result! Furthermore, you will also save much time. While it may take a while to set up your scripts, once they are in place, you can let the computer analyze all of your data while you're having a cup of coffee. This level of automation is what we are striving for throughout the book, and the shell is the centerpiece of our approach.

Third, text is the rule: If your data are stored in a text file, they can be read and written by any machine, and without the need for sophisticated (and expensive) proprietary software. Text files are (and always will be) supported by any operating system and you will still be able to access your data decades from today (while this is not the case for most proprietary file formats). The text-based nature of Unix might seem unusual at first, especially if you are used to graphical interfaces and proprietary software. However, remember that Unix has been around since the early 1970s and will likely be around at the end of your career. Thus, the hard work you put into learning Unix will pay off over a lifetime.

The long history of Unix means that a large body of tutorials and support websites are readily available online. Last but not least, Unix is very stable, robust, secure, and—in the case of Linux—freely available.

In the end, it is almost impossible for a professional scientist to entirely avoid working in a Unix shell: the majority of high-performance computing platforms (computer clusters, large workstations, etc.) run a Unix or Linux operating system. Similarly, the transfer of large files, websites, and data between machines is often accomplished through command-line interfaces.

Mastering the skills presented in this chapter will allow you to work with large files (or with many files) effortlessly. Most operations can be accomplished without the need to open the file(s) in an editor, and can be automated very easily.

1.3 Getting Started with Unix

1.3.1 Installation

The Linux distribution Ubuntu and Apple's OS X are members of the *nix family of operating systems. If you are using either of them, you do not need to install any specific software to follow the material in this chapter.

Microsoft Windows is not based on a *nix system; you can, however, recreate a Unix environment within Windows by installing the Ubuntu operating system in a virtual machine. Alternatively, Windows users can install Git Bash, a clone of the Bash terminal. It provides basic Unix and Git commands and many other standard Unix functionalities can be installed.

Please find instructions for its installation in CSB/unix/installation. Windows also ships with the program Command Prompt, a command-line interface for Windows. However, many commands differ from their Bash shell counterparts, so we will not cover these here.

1.3.2 Directory Structure

In Unix we speak of "directories," while in a graphical environment the term "folder" is more common. These two terms are interchangeable and refer to a structure that may contain subdirectories and files. The Unix directory structure is organized hierarchically in a tree. Figure 1.1 illustrates a directory structure in the OS X operating system. The topmost directory in the hierarchy is also called the "root" directory and is denoted by an individual slash (/). The precise architecture varies among the different operating systems, but there are some important directories that branch off the root directory in most operating systems:

/bin   Contains several basic programs
/dev   Contains the files connecting to devices such as the keyboard, mouse, and screen
/etc   Contains configuration files
/tmp   Contains temporary files

Another important directory is your home directory (also called the login directory), which is the starting directory when you open a new shell. It contains your personal files, directories, and programs. The tilde (~) symbol is shorthand for the home directory in Ubuntu and OS X. The exact path to your home directory varies slightly among different operating systems. To print its location, open a terminal and type[1]

```
echo $HOME
```

The command echo prints a string to the screen. The dollar sign indicates a variable. You will learn more about variables in section 1.7.

If you followed the instructions in section 0.5, you should have created a directory called CSB in your home directory. In Ubuntu the location is /home/YOURNAME/CSB, in OS X it is /Users/YOURNAME/CSB. Windows users need

---

1. Windows users, use Git Bash or type echo %USERPROFILE% at the Windows Command Prompt.

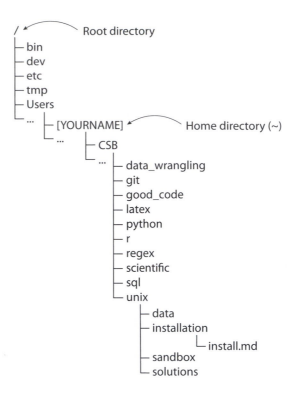

Full path of the file install.md:
```
/Users/[YOURNAME]/CSB/unix/installation/install.md
```

Figure 1.1. An example of the directory structure in the OS X operating system. It shows several directories branching off the root directory (/). In OS X the home directory (~) is a subdirectory of Users, and in UNIX it is a subdirectory of home. If you have followed the instructions in section 0.5, you will find the directory CSB in your home directory. As an example, we show the full path of the file install.md.

to decide where to store the directory. Within CSB, you will find several directories, one for each chapter (e.g., CSB/unix). Each of these directories contains the following subdirectories:

installation The instructions for installing the software needed for the chapter are contained here. These are also available online.[2]

sandbox This is the directory where we work and experiment.

data This directory provides all data for the examples and exercises, along with the corresponding citations for papers and websites.

---

2. github.com/CSB-book/CSB.

solutions   The detailed solutions for the exercises are here, as well as sketches of the solutions in plain English (*pseudocode*) that you should consult if you don't know how to proceed with an exercise. Solutions for the "Intermezzo" sections are available at the end of the book.

When you navigate the system, you are in one directory and can move deeper in the tree, or upward toward the root. Section 1.4.3 discusses the commands you can use to move between the hierarchical levels and determine your location within the directory structure.

1.4 Getting Started with the Shell

In Ubuntu, you can open a shell by pressing Ctrl+Alt+T, or by opening the dash (hold the Meta key) and typing `Terminal`. In OS X, you want to open the application `Terminal.app`, which is located in the folder `Utilities` within `Applications`. Alternatively, you can type "Terminal" in `Spotlight`. Windows users can launch Git Bash or another terminal emulator. In all systems, the shell automatically starts in your home directory.

When you open a terminal, you should see a line (potentially containing information on your user name and location), ending with a dollar ($) sign. When you see the dollar sign, the terminal is ready to accept your commands. Give it a try and type

```
# display date and time
$ date
```

In this book, a $ sign at the beginning of a line of code signals that the command has to be executed in your terminal. You do not need to type the $ sign in your terminal, only copy the command that follows it. A line starting with a hash symbol (#) means that everything that follows is a comment. While Unix ignores comments, you will find hints, reminders, and explanations there. Make plenty of use of comments to document your own code. When writing multiple lines of comments, start each with #.

In Unix, you can use the Tab key to reduce the amount you have to type, which in turn reduces the probability of making mistakes. When you press Tab in a (properly configured) shell, it will try to automatically complete your command, directory, or file name. If multiple completions are possible, you can display them all by hitting the Tab key twice. Additionally, you can navigate the history of commands you have typed by using the up/down arrows. This is very convenient as you do not need to retype a command that you have executed recently. The following box lists keyboard shortcuts that help pace through long lines of code.

Ctrl+A	Go to the beginning of the line.
Ctrl+E	Go to the end of the line.
Ctrl+L	Clear the screen.
Ctrl+U	Clear the line before the cursor position.
Ctrl+K	Clear the line after the cursor.
Ctrl+C	Kill the command that is currently running.
Ctrl+D	Exit the current shell.
Alt+F	Move cursor forward one word (in OS X, Esc+F).
Alt+B	Move cursor backward one word (in OS X, Esc+B).

Mastering these and other keyboard shortcuts will save you a lot of time. You may want to print this list (available at computingskillsfor biologists.com/terminalshortcuts) and keep it next to your keyboard—when used consistently you will have them all memorized and will start using them automatically.

1.4.1 Invoking and Controlling Basic Unix Commands

Some commands can be executed simply by typing their name:

```
# print a simple calendar
$ cal
```

However, you can pass an *argument* to the command to alter its behavior:

```
# pass argument to cal to print specific year
$ cal 2020
```

Some commands may require obligatory arguments and will return an error message if they are missing. For example, the command to copy a file needs two arguments: what file to copy, and where to copy it (see section 1.5.1).

In addition, all commands can be modified using *options* that are specific to each command. For instance, we can print the calendar in Julian format, which labels each day with a number starting on January 1:

```
# use option -j to display Julian calendar
$ cal -j
```

Options can be written using either a dash followed by a single letter (older style, e.g., -j) or two dashes followed by words (newer style, e.g., --julian). Note that not every command offers both styles.

In Unix, the placement of spaces between a command and its options or arguments is important. There needs to be a space between the command and its options, and between multiple arguments. However, if you are supplying multiple options to a command you can string them together (e.g., -xvzf).

If you are using Unix commands for the first time, it might seem odd that you usually do not get a message or response after executing a command (unless the command itself prints to the screen). Some commands provide feedback on their activity when you request it using an option. Otherwise, seeing no (error) message means that your command worked.

Last but not least, it is important to know how to interrupt the execution of a command. Press Ctrl+C to halt the command that is currently running in your shell.

1.4.2 How to Get Help in Unix

Unix ships with hundreds of commands. As such, it is impossible to remember them all, let alone all their possible options. Fortunately, each command is described in detail in its manual page, which OS X and Ubuntu users can access directly from the shell by typing man [COMMAND]. Use arrows to scroll up and down and press q to close the manual page. Users of Git Bash can search online for unix man page [COMMAND] to find many sites displaying the manual of Unix commands.

Checking the exact behavior of a command is especially important, given that the shell will execute any command you type without asking whether you know what you're doing (so that it will promptly remove all of your files, if that's the command you typed). You may be used to more forgiving (and slightly patronizing) operating systems in which a pop-up window will warn you whenever something you're doing is considered dangerous. However, don't feel afraid to use the shell as much as possible. The really dangerous commands are all very specific—there is very little chance that you will destroy something accidentally simply by hitting the wrong key.

1.4.3 Navigating the Directory System

You can navigate the hierarchical Unix directory system using the following commands:

cd   <u>C</u>hange <u>d</u>irectory. The command requires one argument: the path to the directory you want to change into. There are a few options for the command that speed up navigation through the directory structure:

cd ..   Move one directory up.
cd /   Move to the root directory.
cd ~   Move to your home directory.
cd -   Go back to the directory you visited previously (like "Back" in a browser).

```
# assuming you saved CSB in your home directory
# navigate to the sandbox in the CSB/unix directory
cd ~/CSB/unix/sandbox
```

pwd   <u>P</u>rint the path of the current <u>w</u>orking <u>d</u>irectory. This command prints your current location within the directory structure.

```
$ pwd
/Users/mwilmes/CSB/unix/sandbox
# this may look different depending on your system
```

ls   <u>L</u>ist the files and subdirectories in the current directory. There are several useful options:

ls -a   List <u>a</u>ll files (including hidden files).

ls -l   Return a long list with details on access permissions (see section 1.6.7), the number of links to a file, the user and group owning it, its size, the time and date it was changed last, and its name.

ls -lh   Display file sizes in human readable units (K, M, G for kilobytes, megabytes, gigabytes, respectively).

One can navigate through the directory hierarchy by providing either the *absolute* path or a *relative* path. An example of the absolute path of a directory is indicated at the bottom of figure 1.1. The full path of the file install.md is indicated, starting at the root. A relative path is defined with respect to the current directory (use pwd to display the absolute path of your current working directory). Let's look at an example:

```
# absolute path to the directory CSB/python/data
# if CSB is in your home directory
$ cd ~/CSB/python/data
# relative path to navigate to CSB/unix/data
# remember: the Tab key provides autocomplete
$ cd ../../unix/data
# go back to previous directory (CSB/python/data)
$ cd -
```

You can always use either the absolute path or a relative path to specify a directory. When you navigate just a few levels higher or deeper within the tree, a relative path usually means less to type. If you want to jump somewhere far within the tree, the absolute path might be the better choice.

Note that directory names in a path are separated with a forward slash (/) in Unix but usually with a backslash (\) in Windows (e.g., when you look at the path of a file using File Explorer). However, given that Git Bash emulates a Unix environment, you will find it uses a forward slash despite working with the Windows operating system.

In Unix, a full path name cannot have any spaces. Spaces in your file or directory names need to be preceded by a backslash (\). For example, the file "My Manuscript.txt" in the directory "Papers and reviews" becomes Papers\ and\ reviews/My\ Manuscript.txt. To avoid such unruly path names, an underscore (_) is recommended for separating elements in the names of files and directories, rather than a space. If you need to refer to an existing file or directory that has spaces in its name, use quotation marks around it.

```
# does not work
cd Papers and reviews
# works but not optimal
cd Papers\ and\ reviews
cd "Papers and reviews"
# when creating files or directories use
# underscores to separate elements in their names
cd Papers_and_reviews
```

Intermezzo 1.1

(a) Go to your home directory.
(b) Navigate to the sandbox directory within the CSB/unix directory.
(c) Use a relative path to go to the data directory within the python directory.
(d) Use an absolute path to go to the sandbox directory within python.
(e) Return to the data directory within the python directory.

1.5 Basic Unix Commands

1.5.1 Handling Directories and Files

Creating, manipulating and deleting files or directories is one of the most common tasks you will perform. Here are a few useful commands:

cp   Copy a file or directory. The command requires two arguments: the first argument is the file or directory that you want to copy, and the second is the location to which you want to copy. In order to copy a directory, you need to add the option -r which makes the command *recursive*. The directory and its contents, including subdirectories (and their contents) will be copied.

```
# copy a file from unix/data directory into sandbox
# if you specify the full path,
# your current location does not matter
$ cp ~/CSB/unix/data/Buzzard2015_about.txt ~/CSB/unix/
    ↳ sandbox/
# assuming your current location is the unix sandbox,
```

```
# we can use a relative path
$ cp ../data/Buzzard2015_about.txt .
# the dot is shorthand to say "here"
# rename the file in the copying process
cp ../data/Buzzard2015_about.txt ./Buzzard2015_about2.txt
# copy a directory (including all subdirectories)
cp -r ../data .
```

mv   Move or rename a file or directory. You can move a file by specifying two arguments: the name of the file or directory you want to move, and the destination. You can also use the mv command to rename a file or directory. Simply specify the old and the new file name in the same location.

```
# move the file to the data directory
$ mv Buzzard2015_about2.txt ../data/
# rename a file
$ mv ../data/Buzzard2015_about2.txt ../data/
   ↳ Buzzard2015_about_new.txt
# easily manipulate a file that is
# not in your current working directory
```

touch   Update the date of last access to the file. Interestingly, if the file does not exist, this command will create an empty file.

```
# inspect the current contents of the directory
$ ls -l
# create a new file (you can list multiple files)
$ touch new_file.txt
# inspect the contents of the directory again
$ ls -l
# if you touch the file a second time,
# the time of last access will change
```

rm   Remove a file. It has some useful options: rm -r deletes the contents of a directory recursively (i.e., including all files and subdirectories in it). Use this command with caution or in conjunction with the -i option, which prompts the user to confirm the action. The option -f forcefully removes a write-protected file (such as a directory under version control) without a prompt.

Again, use with caution as there is no trash bin that allows you to undo the removal!

```
$ rm -i new_file.txt
remove new_file.txt? y
# confirm deletion with y (yes) or n (no)
```

mkdir   Make a directory. To create nested directories, use the option -p:

```
$ mkdir -p d1/d2/d3
# remove the directory by using command rm recursively
$ rm -r d1
```

1.5.2 Viewing and Processing Text Files

Unix was especially designed to handle text files, which is apparent when considering the multitude of commands dealing with text. Here are a few popular ones with a selection of useful options:[3]

less   Progressively print a file to the screen. With this command you can instantly take a look at very large files without the need to open them. In fact, less does not load the entire file, but only what needs to be displayed—making it much faster than a text editor. Once you enter the less environment, you have many options to navigate through the file or search for specific patterns. The simplest are Ctrl+F to jump one screen forward and Ctrl+B to jump one back. See more options by pressing h or have a look at the manual page. Pressing q quits the less environment.[4]

```
# assuming you're in CSB/unix/data
$ less Marra2014_data.fasta
>contig00001  length=527  numreads=2  gene=isogroup00001
    ↳ status=it_thresh
```

---

3. We recommend skimming the manual pages of each command to get a sense of their full capabilities.

4. Funny fact: there is a command called more that does the same thing, but with less flexibility. Clearly, in Unix, less is more.

```
ATCCTAGCTACTCTGGAGACTGAGGATTGAAGTTCAAAGTCAGCTCAAGCAAGAGATTT
...
```

cat   Concatenate and print files. The command requires at least one file
name as argument. If you provide only one, it will simply print the entire file
contents to the screen. Providing several files will concatenate the contents of
all files and print them to the screen.

```
# concatenate files and print to screen
$ cat Marra2014_about.txt Gesquiere2011_about.txt
    ↳ Buzzard2015_about.txt
```

wc   Line, word, and byte (character) count of a file. The option -l returns the
line count only and is a quick way to get an idea of the size of a text file.

```
# count lines, words, and characters
$ wc Gesquiere2011_about.txt
      8      64     447 Gesquiere2011_about.txt
# count lines only
$ wc -l Marra2014_about.txt
     14 Marra2014_about.txt
```

sort   Sort the lines of a file and print the result to the screen. Use option -n
for numerical sorting and -r to reverse the order. The option -k is useful to
sort a delimiter-separated file by a specific column (more on this command
in section 1.6.3).

```
# print the sorted lines of a file
$ sort Gesquiere2011_data.csv
100     102.56  163.06
100     117.05  158.01
100     133.4   94.78
...
# sort numerically
$ sort -n Gesquiere2011_data.csv
```

```
maleID  GC       T
1       32.65    59.94
1       51.09    35.57
1       52.72    43.98
...
```

uniq   Show only the <u>uniq</u>ue lines of a file. The contents need to be sorted first for this to work properly. Section 1.6.2 describes how to combine commands. The option -c returns a count of occurrence for each unique element in the input.

file   Determine the type of a <u>file</u>. Useful to identify Windows-style line terminators[5] before opening a file.

```
$ file Marra2014_about.txt
Marra2014_about.txt: ASCII English text
```

head   Print the <u>head</u> (i.e., first few lines) of a file. The option -n determines the number of lines to print.

```
# display first two lines of a file
head -n 2 Gesquiere2011_data.csv
maleID  GC       T
1       66.9     64.57
```

tail   Print the <u>tail</u> (i.e., last few lines) of a file. The option -n controls the number of lines to print (starting from the end of the file). The option can also be used to display everything but the first few lines.

```
# display last two lines of file
$ tail -n 2 Gesquiere2011_data.csv
127     108.08   152.61
127     114.09   151.07
```

---

5. Covered in section 1.9.2.

```
# display from line 2 onward
# (i.e., removing the header of the file)
$ tail -n +2 Gesquiere2011_data.csv
1       66.9    64.57
1       51.09   35.57
1       65.89   114.28
...
```

diff   Show the differences between two files.

Intermezzo 1.2

To familiarize yourself with these basic Unix commands, try the following:

(a) Go to the data directory within CSB/unix.
(b) How many lines are in file Marra2014_data.fasta?
(c) Create the empty file toremove.txt in the CSB/unix/sandbox directory without leaving the current directory.
(d) List the contents of the directory unix/sandbox.
(e) Remove the file toremove.txt.

1.6 Advanced Unix Commands

1.6.1 Redirection and Pipes

So far, we have printed the output of each command (e.g., ls) directly to the screen. However, it is easy to *redirect* the output to a file or to *pipe* the output of one command as the input to another command. Stringing commands together using pipes is the real power of Unix, letting you perform complex tasks on large amounts of data using a single line of commands.

First, we show how to redirect the output of a command into a file:

```
$ [COMMAND] > filename
```

Note that if the file filename exists, it will be overwritten. If instead we want to append the output of a command to an existing file, we can use the >> symbol, as in the following line:

```
$ [COMMAND] >> filename
```

When the command is very long and complex, we might want to redirect the contents of a file as input to a command, "reversing" the flow:

```
$ [COMMAND] < filename
```

To run a few examples, let's start by moving to our sandbox:

```
$ cd ~/CSB/unix/sandbox
```

The command echo can be used to print a string on the screen. Instead of printing to the screen, we redirect the output to a file, effectively creating a file containing the string we want to print:

```
$ echo "My first line" > test.txt
```

We can see the result of our operation by printing the file to the screen using the command cat:

```
$ cat test.txt
```

To append a second line to the file, we use >>:

```
$ echo "My second line" >> test.txt
$ cat test.txt
```

We can redirect the output of any command to a file.

Here is an example: Your collaborator or laboratory machine provided you with a large number of data files. Before analyzing the data, you want to get a sense of how many files need to be processed. If there are thousands of files, you wouldn't want to count them manually or even open a file browser that could do the counting for you. It is much simpler and faster to type a few Unix commands.

We will use unix/data/Saavedra2013 as an example of a directory with many files. First, we create a file that lists all the files contained in the directory:

```
# current directory is the unix sandbox
# create a file listing the contents of a directory
$ ls ../data/Saavedra2013 > filelist.txt
# look at the file
$ cat filelist.txt
```

Now we want to count how many lines are in the file. We can do so by calling the command wc -l:[6]

```
# count lines in a file
$ wc -l filelist.txt
# remove the file
$ rm filelist.txt
```

However, we can skip the creation of the intermediate file (filelist.txt) by creating a short pipeline. The pipe symbol (|) tells the shell to take the output on the left of the pipe and use it as the input for the command on the right of the pipe. To take the output of the command ls and use it as the input of the command wc we can write

```
# count number of files in a directory
$ ls ../data/Saavedra2013 | wc -l
```

We have created our first, simple pipeline. In the following sections, we are going to build increasingly long and complex pipelines. The idea is always to start with a command and progressively add one piece after another to the pipeline, each time checking that the result is the desired one.

1.6.2 Selecting Columns Using cut

When dealing with tabular data, you will often encounter the comma-separated values (CSV) standard file format. As the name implies, the data are usually structured by commas, but you may find CSV files using other

---

6. This is the lowercase letter L as in line.

delimiters such as semicolons or tabs (e.g., because the data values contain commas and spaces). The CSV format is text based and platform and software independent, making it the standard output format for many experimental devices. The versatility of the file format should also make it your preferred choice when manually entering and storing data.[7] Most of the exercises in this book use the CSV file format in order to highlight how easy it is to read and write these files using different programming languages.

The main Unix command you want to master for comma-, space-, tab-, or character-delimited text files is cut. To showcase its features, we work with data on the generation time of mammals published by Pacifici et al. (2013). First, let's make sure we are in the right directory (~/CSB/unix/data). Then, we can print the header (the first line, specifying the contents of each column) of the CSV file using the command head, which prints the first few lines of a file on the screen, with the option -n 1, specifying that we want to output only the first line:

```
# change directory
$ cd ~/CSB/unix/data
# display first line of file (i.e., header of CSV file)
$ head -n 1 Pacifici2013_data.csv
TaxID;Order;Family;Genus;Scientific_name;...
```

We can see that the data are separated by semicolons. We pipe the first line of the file to cut and use the option -d ";" to specify the delimiter. The additional option -f lets us extract specific columns: here column 1 (-f 1), or the first four columns (-f 1-4).

```
# take first line, select 1st column of ";"-separated file
$ head -n 1 Pacifici2013_data.csv | cut -d ";" -f 1
TaxID

$ head -n 1 Pacifici2013_data.csv | cut -d ";" -f 1-4
TaxID;Order;Family;Genus
```

Remember to use the Tab key to autocomplete file names and the arrow keys to access your command history.

---

7. If you need to store and process large data sets, you should consider databases, which we explore in chapter 10, as an alternative.

In the next example we work with the contents of our data file. We specify a delimiter, extract specific columns, and pipe the result to the head command in order to display only the first few elements:

```
# select 2nd column, display first 5 elements
$ cut -d ";" -f 2 Pacifici2013_data.csv  | head -n 5
Order
Rodentia
Rodentia
Rodentia
Macroscelidea

# select 2nd and 8th columns, display first 3 elements
$ cut -d ";" -f 2,8 Pacifici2013_data.csv  | head -n 3
Order;Max_longevity_d
Rodentia;292
Rodentia;456.25
```

Now, we specify the delimiter, extract the second column, skip the first line (the header) using the tail -n +2 command (i.e., return the whole file starting from the second line), and finally display the first five entries:

```
# select 2nd column without header, show 5 first elements
$ cut -d ";" -f 2 Pacifici2013_data.csv  | tail -n +2 |
    ↳ head -n 5
Rodentia
Rodentia
Rodentia
Macroscelidea
Rodentia
```

We pipe the result of the previous command to the sort command (which sorts the lines), and then again to uniq (which takes only the elements that are not repeated).[8] Effectively, we have created a pipeline to extract the names of all the orders in the database, from Afrosoricida to Tubulidentata (a remarkable order, which today contains only the aardvark).

---

8. The command uniq is typically used in conjunction with sort, as it will remove duplicate lines only if they are contiguous.

```
# select 2nd column without header, unique sorted elements
$ cut -d ";" -f 2 Pacifici2013_data.csv  | tail -n +2 |
  ↳ sort | uniq
Afrosoricida
Carnivora
Cetartiodactyla
...
```

This type of manipulation of character-delimited files is very fast and effective. It is an excellent idea to master the cut command in order to start exploring large data sets without the need to open files in specialized programs. (Note that opening a file in a text editor might modify the contents of a file without your knowledge. Find details in section 1.9.2.)

Intermezzo 1.3

(a) If we order all species names (fifth column) of Pacifici2013_data.csv in alphabetical order, which is the first species? Which is the last?

(b) How many families are represented in the database?

1.6.3 Substituting Characters Using tr

Often we want to substitute or remove a specific character in a text file (e.g., to convert a comma-separated file into a tab-separated file). Such a one-by-one substitution can be accomplished with the command tr. Let's look at some examples in which we use a pipe to pass a string to tr, which then processes the text input according to the search term and specific options.

Substitute all characters a with b:

```
$ echo "aaaabbb" | tr "a" "b"
bbbbbbb
```

Substitute every digit in the range 1 through 5 with 0:

```
$ echo "123456789" | tr 1-5 0
000006789
```

Substitute lowercase letters with uppercase ones:

```
$ echo "ACtGGcAaTT" | tr actg ACTG
ACTGGCAATT
```

We obtain the same result by using bracketed expressions that provide a predefined set of characters. Here, we use the set of all lowercase letters [:lower:] and translate into uppercase letters [:upper:]:

```
$ echo "ACtGGcAaTT" | tr [:lower:] [:upper:]
ACTGGCAATT
```

We can also indicate ranges of characters to substitute:

```
$ echo "aabbccddee" | tr a-c 1-3
112233ddee
```

Delete all occurrences of a:

```
$ echo "aaaaabbbb" | tr -d a
bbbb
```

"Squeeze" all consecutive occurrences of a:

```
$ echo "aaaaabbbb" | tr -s a
abbbb
```

Note that the command tr cannot operate on a file "in place," meaning that it cannot change a file directly. However, it can operate on a copy of the contents of a file. For instance, we can use pipes in conjunction with cat, head, cut, or the output redirection operator to create input for tr:

```
# pipe output of cat to tr
$ cat inputfile.csv | tr " " "\t" > outputfile.csv
# redirect file contents to tr
$ tr " " "\t" < inputfile.csv > outputfile.csv
```

In this example we replace all spaces within the file inputfile.csv with tabs. Note the use of quotes to specify the space character. The tab is indicated by \t. The backslash defines a *metacharacter*: it signals that the following character should not be interpreted literally, but rather represents a special code referring to a character (e.g., a tab) that is difficult to represent otherwise.

Now we can apply the command tr and the commands we showcased earlier to create a new file containing a subset of the data contained in Pacifici2013_data.csv, which we are going to use in the next section.

First, we change directory to the sandbox:

```
$ cd ../sandbox/
```

To recap, we were working in the directory ~/CSB/unix/data. We then moved one directory up (..) to get to the directory ~/CSB/unix/, from which we moved down into the sandbox.

Now we want to create a version of Pacifici2013_data.csv containing only the Order, Family, Genus, Scientific_name, and AdultBodyMass_g (columns 2–6). Moreover, we want to remove the header, sort the lines according to body mass (with larger critters first), and have the values separated by spaces. This sounds like an awful lot of work, but we're going to see how this can be accomplished by piping a few commands together.

First, let's remove the header:

```
$ tail -n +2 ../data/Pacifici2013_data.csv
```

Then, take only columns 2–6:

```
$ tail -n +2 ../data/Pacifici2013_data.csv | cut -d ";"
  ↳ -f 2-6
```

Now, substitute the current delimiter (;) with a space:

```
$ tail -n +2 ../data/Pacifici2013_data.csv | cut -d ";"
  ↳ -f 2-6 | tr ";" " "
```

To sort the lines according to body size, we need to exploit a few of the options for the command sort. First, we want to sort numbers (option -n);

second, we want larger values first (option -r, reverse order); finally, we want to sort the data according to the sixth column (option -k 6):

```
$ tail -n +2 ../data/Pacifici2013_data.csv | cut -d ";"
    ↳ -f 2-6 | tr ";" " " | sort -r -n -k 6
```

That's it. We have created our first complex pipeline. To complete the task, we redirect the output of our pipeline to a new file called BodyM.csv:

```
$ tail -n +2 ../data/Pacifici2013_data.csv | cut -d ";"
    ↳ -f 2-6 | tr ";" " " | sort -r -n -k 6 > BodyM.csv
```

You might object that the same operations could have been accomplished with a few clicks by opening the file in a spreadsheet editor. However, suppose you have to repeat this task many times; for example, you have to reformat every file that is produced by a laboratory device. Then it is convenient to automate this task such that it can be run with a single command. This is exactly what we are going to do in section 1.7.

Similarly, suppose you need to download a large CSV file from a server, but many of the columns are not needed. With cut, you can extract just the relevant columns, reducing download time and storage.

1.6.4 Wildcards

Wildcards are special symbols that work as placeholders for one or more characters. The *star wildcard* (*) stands for zero or more characters with the exception of a leading dot. Unix uses a leading dot for hidden files, so this means that hidden files are ignored in a search using this wildcard (show hidden files using ls -a). A question mark (?) is a placeholder for any single character, again with the exception of a leading dot.

Let's look at some examples in the directory CSB/unix/data/miRNA:

```
# change into the directory
$ cd ~/CSB/unix/data/miRNA
# count the numbers of lines in all the .fasta files
```

```
$ wc -l *.fasta
     714 ggo_miR.fasta
    5176 hsa_miR.fasta
     166 ppa_miR.fasta
    1320 ppy_miR.fasta
    1174 ptr_miR.fasta
      20 ssy_miR.fasta
    8570 total
# print the first two lines of each file
# whose name starts with pp
$ head -n 2 pp*
==> ppa_miR.fasta <==
>ppa-miR-15a MIMAT0002646
UAGCAGCACAUAAUGGUUUGUG

==> ppy_miR.fasta <==
>ppy-miR-569 MIMAT0016013
AGUUAAUGAAUCCUGGAAAGU


# determine the type of every file that has
# an extension with exactly three letters
$ file *.???
```

1.6.5 Selecting Lines Using grep

grep is a powerful command that finds all the lines of a file that match a given pattern. You can return or count all occurrences of the pattern in a large text file without ever opening it. grep is based on the concept of regular expressions, which we will cover in depth in chapter 5.

We explore the basic features of grep using the file we created in section 1.6.3. The file contains data on thousands of species:

```
$ cd ~/CSB/unix/sandbox
$ wc -l BodyM.csv
5426 BodyM.csv
```

Let's see how many wombats (family Vombatidae) are contained in the data. To display the lines that contain the term "Vombatidae" we execute grep with two arguments—the search term and the file that we want to search in:

```
$ grep Vombatidae BodyM.csv
Diprotodontia Vombatidae Lasiorhinus Lasiorhinus krefftii
    ↳ 31849.99
Diprotodontia Vombatidae Lasiorhinus Lasiorhinus latifrons
    ↳ 26163.8
Diprotodontia Vombatidae Vombatus Vombatus ursinus 26000
```

Now we add the option -c to count the lines that contain a match:

```
$ grep -c Vombatidae BodyM.csv
3
```

Next, we have a look at the genus *Bos* in the data file:

```
$ grep Bos BodyM.csv
Cetartiodactyla Bovidae Bos Bos sauveli 791321.8
Cetartiodactyla Bovidae Bos Bos gaurus 721000
Cetartiodactyla Bovidae Bos Bos mutus 650000
Cetartiodactyla Bovidae Bos Bos javanicus 635974.3
Cetartiodactyla Bovidae Boselaphus Boselaphus tragocamelus
    ↳ 182253
```

Besides all the members of the *Bos* genus, we also match one member of the genus *Boselaphus*. To exclude it, we can use the option -w, which prompts grep to match only full words:

```
$ grep -w Bos BodyM.csv
Cetartiodactyla Bovidae Bos Bos sauveli 791321.8
Cetartiodactyla Bovidae Bos Bos gaurus 721000
Cetartiodactyla Bovidae Bos Bos mutus 650000
Cetartiodactyla Bovidae Bos Bos javanicus 635974.3
```

Using the option -i we can make the search case insensitive (it will match both upper- and lowercase instances):

```
$ grep -i Bos BodyM.csv
Proboscidea Elephantidae Loxodonta Loxodonta africana
    ↳ 3824540
```

```
Proboscidea Elephantidae Elephas Elephas maximus 3269794
Cetartiodactyla Bovidae Bos Bos sauveli 791321.8
Cetartiodactyla Bovidae Bos Bos gaurus 721000
...
```

Sometimes, we want to know which lines precede or follow the one we want to match. For example, suppose we want to know which mammals have body weight most similar to the gorilla (*Gorilla gorilla*). The species are already ordered by size (see section 1.6.3), thus we can simply print the two lines before the match using the option -B 2 and the two lines after the match using -A 2:

```
$ grep -B 2 -A 2 "Gorilla gorilla" BodyM.csv
Cetartiodactyla Bovidae Ovis Ovis ammon 113998.7
Cetartiodactyla Delphinidae Lissodelphis Lissodelphis
    ↳ borealis 113000
Primates Hominidae Gorilla Gorilla gorilla 112589
Cetartiodactyla Cervidae Blastocerus Blastocerus
    ↳ dichotomus 112518.5
Cetartiodactyla Iniidae Lipotes Lipotes vexillifer
    ↳ 112138.3
```

Use option -n to show the line number of the match. For example, the gorilla is the 164th largest mammal in the database:

```
$ grep -n "Gorilla gorilla" BodyM.csv
164:Primates Hominidae Gorilla Gorilla gorilla 112589
```

To print all the lines that do not match a given pattern, use the option -v. For instance, we want to find species of the genus *Gorilla* other than *Gorilla gorilla*. We can pipe the result of matching all members of the genus *Gorilla* to a second grep statement that excludes the species *Gorilla gorilla*:

```
$ grep Gorilla BodyM.csv | grep -v gorilla
Primates Hominidae Gorilla Gorilla beringei 149325.2
```

To match one of several strings, use grep "[STRING1]\|[STRING2]":

```
$ grep -w "Gorilla\|Pan" BodyM.csv
Primates Hominidae Gorilla Gorilla beringei 149325.2
Primates Hominidae Gorilla Gorilla gorilla 112589
Primates Hominidae Pan Pan troglodytes 45000
Primates Hominidae Pan Pan paniscus 35119.95
```

You can use grep on multiple files at a time! Simply list all the files that you want to search (or use wildcards to specify multiple file names). Finally, use the recursive search option -r to search for patterns within all the files in a directory. For example,

```
$ cd ~/CSB/unix
# search recursively in the data directory
$ grep -r "Gorilla" data
```

1.6.6 Finding Files with find

The find command is the command-line program to locate files in your system. You can search by file name, owner, group, type, and other criteria. For example, find the files and subdirectories that are contained in the unix/data directory:

```
# current directory is the unix sandbox
$ find ../data
../data
../data/Buzzard2015_data.csv
../data/Pacifici2013_data.csv
../data/Gesquiere2011_about.txt
../data/Gesquiere2011_data.csv
../data/Saavedra2013
...
```

To count occurrences, we can pipe to wc -l:

```
$ find ../data | wc -l
77
```

Now we can use find to match particular files. First, we specify where to search: this could be either an absolute path (e.g., /home/YOURNAME/CSB/unix/data) or a relative one (e.g., ../data, provided we're in unix/sandbox).

If we want to match a specific file name, we can use the option -name:

```
$ find ../data -name "n30.txt"
../data/Saavedra2013/n30.txt
```

To exploit the full power of find, we use wildcards.[9] For example, use the * wildcard to find all the files whose names contain the word about; the option -iname ignores the case of the file name:

```
$ find ../data -iname "*about*"
../data/Gesquiere2011_about.txt
../data/Buzzard2015_about.txt
...
```

You can specify the depth of the search by limiting it to, for example, only the directories immediately descending from the current one. See the difference between

```
$ find ../data -name "*.txt" | wc -l
64 # depending on your system
```

and

```
$ find ../data -maxdepth 1 -name "*.txt" | wc -l
5
```

which excluded all files in subdirectories. You can exclude certain files:

```
$ find ../data -not -name "*about*" | wc -l
72
```

---

9. See section 1.6.4 for an introduction to wildcards.

or find only directories:

```
$ find ../data -type d
../data
../data/miRNA
../data/Saavedra2013
```

Intermezzo 1.4

    (a) Navigate to CSB/unix/sandbox. Without navigating to a different location, find a CSV file that contains Dalziel in its file name and is located within the CSB directory. Copy this file to the Unix sandbox.

    (b) Print the first few lines on the screen to check the structure of the data. List all unique cities in column loc (omit the header). How often does each city occur in the data set?

    (c) The fourth column reports cases of measles. What is the maximum number of cases reported for Washington, DC?

    (d) What is the maximum number of reported measles cases in the entire data set? Where did this occur?

1.6.7 Permissions

In Unix, each file and directory has specific security attributes specifying who can read (r), write (w), execute (x), or do nothing (-) with the file or directory. These permissions are specified for three entities that may wish to manipulate the file (owner, specific group, and others). The group level is useful for assigning permissions to a specific group of users (e.g., administrators, developers) but not everyone else.

Typing ls -l lists the permissions of each file or subdirectory at the beginning of the line. Each permission is represented by a 10-character notation. The first character refers to the file type and is not related to permissions (- means file, d stands for directory). The last 9 are arranged in groups of 3 (triads) representing the ownership groups (owner, group, others). For example, when a file has the permission -rwxr-xr--, the owner of this file can read, write, and execute the file (rwx), the group can read and execute (r-x), while everyone else can only read (r--).

The commands chmod and chown change the permissions and ownership of a file, respectively:

```
# create a file in the unix sandbox
$ touch permissions.txt
# look at the current permissions
# (output will be different on your machine)
$ ls -l
-rw-r--r--  1 mwilmes  staff  0 Aug 15 09:47 permissions.
    ↳ txt
# change permissions (no spaces between mode settings)
$ chmod u=rwx,g=rx,o=r permissions.txt
# look at changes in permissions
$ ls -l
-rwxr-xr--  1 mwilmes  staff  0 Aug 15 09:48 permissions.
    ↳ txt
# take execute permission away from group,
# and add write rights to others
$ chmod g-x,o+w permissions.txt
$ ls -l
-rwxr--rw-  1 mwilmes  staff  0 Aug 15 09:49 permissions.
    ↳ txt
```

Some operations, such as changing the ownership of a file or directory, or installing new software, can be performed only by the administrator of the machine. You, however, can do so by typing the word sudo (substitute user do) in front of the command. The system will request a password and, if you are authorized to use the sudo command, you grant yourself administrator rights.

When you download and install new software, the system will often request the administrator's password. Pay attention to the trustworthiness of the source of the software before confirming the installation, as you may otherwise install malicious software.

Here is an example of changing the file permissions for a directory recursively (i.e., for all subdirectories and files):

```
# create a directory with a subdirectory
$ mkdir -p test_dir/test_subdir
# look at permissions
$ ls -l
drwxr-xr-x  3 mwilmes  staff  102 Aug 15 10:59 test_dir
# change owner of directory recursively using -R
```

```
$ sudo chown -R sallesina test_dir/
# check for new ownership
$ ls -l
drwxr-xr-x  3 sallesina  staff  102 Aug 15 11:01 test_dir
```

1.7 Basic Scripting

Once a pipeline is in place, it is easy to turn it into a *script*. A script is a text file containing a list of several commands. The commands are then executed one after the other, going through the pipeline in an automated manner. To illustrate the concept, we are going to turn the pipeline in section 1.6.3 into a script.

First, we need to create a file for our Unix script that we can edit using a text editor. The typical extension for a file with shell commands is .sh. In this example, we want to create the file ExtractBodyM.sh, which we can open using our favorite text editor. Create an empty file, either in the editor, or using touch:

```
$ touch ExtractBodyM.sh
```

Open the file in a text editor. In Ubuntu you can use, for example, gedit:

```
$ gedit ExtractBodyM.sh &
```

In OS X, calling open will open the file with the default text editor:[10]

```
$ open ExtractBodyM.sh &
```

The "ampersand" (&) at the end of the line prompts the terminal to open the editor in the background, so that you can still use the same shell while working on the file. Windows users can use any text editor.[11]

Now copy the pipeline that we built throughout the previous sections into the file ExtractBodyM.sh. For now, make sure that it is one long line:

---

10. Use option -a to choose a specific editor (e.g., open -a emacs ExtractBodyM.sh &).

11. Make sure, however, that the editor can save files with the Unix line terminator (LF), otherwise the scripts will not work correctly (details in section 1.9.2). Here's a list of suitable editors: computingskillsforbiologists.com/texteditors.

```
tail -n +2 ../data/Pacifici2013_data.csv | cut -d ";"
    -f 2-6 | tr ";" " " | sort -r -n -k 6 > BodyM.csv
```

and save the file. To run the script, call the command bash and the file name:

```
$ bash ExtractBodyM.sh
```

It is a great idea to immediately write comments for the script, to help you remember what the code does. You can add comments using the hash symbol (#):

```
  # take a CSV file delimited by ";"
2 # remove the header
  # make space separated
  # sort according to the 6th (numeric) column
5 # in descending order
  # redirect to a file
  tail -n +2 ../data/Pacifici2013_data.csv | cut -d ";"
      -f 2-6 | tr ";" " " | sort -r -n -k 6 > BodyM.csv
```

As it stands, this script is very specific: both the input file and the output file names are fixed (*hard coded*). It would be better to leave these names to be decided by the user so that the script can be called for any file with the same format. This is easy to accomplish within the Bash shell: simply use generic arguments (i.e., variables), indicated by the dollar sign ($), followed by the variable name (without a space). Here, we use the number of the argument as the variable name. When the script is run, the generic arguments within the script are replaced by the specific argument that the user supplied when executing the script.

Let's change our script ExtractBodyM.sh accordingly:

```
  # take a CSV file delimited by ";" (first argument)
  # remove the header
3 # make space separated
```

```
     # sort according to the 6th (numeric) column
     # in descending order
6    # redirect to a file (second argument)
     tail -n +2 $1 | cut -d ";" -f 2-6 | tr ";" " " | sort -r -n
        -k 6 > $2
```

The file name (i.e., ../data/Pacifici2013_data.csv) and the result file (i.e., BodyM.csv) have been replaced by $1 and $2, respectively. Now you can launch the modified script from the command line by specifying the input and output files as arguments:

```
$ bash ExtractBodyM.sh ../data/Pacifici2013_data.csv BodyM
   ↳ .csv
```

The final step is to make the script directly executable so that you can skip invoking Bash. We can do so by changing the permissions of the file,

```
$ chmod +rx ExtractBodyM.sh
```

and adding a special line at the beginning of the script telling Unix where to find the program (in this case bash[12]) to execute the script:

```
     #!/bin/bash
2
     # the previous line is not a comment, but a special line
     # telling where to find the program to execute the script;
5    # it should be your first line in all Bash scripts

     # function of script:
8    # take a CSV file delimited by ";" (first argument)
     # remove the header
     # make space separated
```

---

12. If you don't know where the program bash is, you can find out by running whereis bash in your terminal.

```
11    # sort according to the 6th (numeric) column
      # in descending order
      # redirect to a file (second argument)
14    tail -n +2 $1 | cut -d ";" -f 2-6 | tr ";" " " | sort -r -n
          -k 6 > $2
```

Now, this script can be invoked as

```
$ ./ExtractBodyM.sh ../data/Pacifici2013_data.csv BodyM.
    ↳ csv
```

Note the ./ in front of the script's name in order to execute the file.

The long Unix pipe that we constructed over the last few pages can be complicated to read and understand. It is therefore convenient to break it into smaller pieces and save the individual output of each part as a temporary file that can be deleted as a last step in the script:

```
      #!/bin/bash
2     # function of script:
      # take a CSV file delimited by ";" (first argument)
      # remove the header
5     # make space separated
      # sort according to the 6th (numeric) column
      # in descending order
8     # redirect to a file (second argument)

      # remove the header
11    tail -n +2 $1 > $1.tmp1
      # extract columns
      cut -d ";" -f 2-6 $1.tmp1  > $1.tmp2
14    # make space separated
      tr ";" " " < $1.tmp2 > $1.tmp3
      # sort and redirect to output
17    sort -r -n -k 6 $1.tmp3 > $2
      # remove temporary, intermediate files
      rm $1.tmp*
```

This is much more readable, although a little more wasteful, as it creates temporary files only then to delete them. Using intermediate, temporary files, however, allows scripts to be "debugged" easily—just comment the last line out and inspect the temporary files one by one to investigate at which point you obtained an unwanted result.

1.8 Simple for Loops

A for loop allows us to repeat a task with slight variations. For instance, a loop is very useful when you need to perform an identical task on multiple files, or when you want to provide different input arguments for the same command. Instead of writing code for every instance separately, we can use a loop.

As a first example, we want to display the first two lines of all .fasta files in the directory CSB/unix/data/miRNA. We first change the directory and execute the ls command to list its contents:

```
$ cd ~/CSB/unix/data/miRNA
$ ls
ggo_miR.fasta    hsa_miR.fasta    ppa_miR.fasta    ...
```

The directory contains six .fasta files with miRNA sequences of different *Hominidae* species. Now we want to get a quick overview of the contents of the files. Instead of individually calling the head command on each file, we can access multiple files by writing a for loop:

```
$ for file in ggo_miR.fasta hsa_miR.fasta
        do head -n 2 $file
    done
>ggo-miR-31 MIMAT0002381
GGCAAGAUGCUGGCAUAGCUG
>hsa-miR-576-3p MIMAT0004796
AAGAUGUGGAAAAAUUGGAAUC
```

Here we created a variable (file) that stands in for the actual file names that are listed after the in. Instead of listing all files individually after the in, we can also use wildcards to consider all .fasta files in the directory:

```
$ for file in *.fasta
        do head -n 2 $file
    done
```

```
>ggo-miR-31 MIMAT0002381
GGCAAGAUGCUGGCAUAGCUG
>hsa-miR-576-3p MIMAT0004796
AAGAUGUGGAAAAAUUGGAAUC

...
```

The actual statement (i.e., what to do with the variable) is preceded by a do. As shown in section 1.7, the variable is invoked with a $ (dollar sign). The statement ends with done. Instead of this clear coding style that spans multiple lines, you may also encounter loops written in one line, using a ; as command terminator instead of line breaks.

In our second example, we call a command with different input variables. Currently, the files in CSB/unix/data/miRNA provide files that contain different miRNA sequences per species. However, we might need files that contain all sequences of different species per type of miRNA. We can accomplish this by using the command grep in a for loop instead of writing code for every type of miRNA separately:

```
$ for miR in miR-208a miR-564 miR-3170
        do grep $miR -A1 *.fasta > $miR.fasta
    done
```

We have created the variable miR that cycles through every item in the list that is given after the in (i.e., types of miRNA). In every iteration of the loop, one instance of the variable is handed to grep. We used the same variable again to create appropriate file names.

Let's have a look at the head of one of the files that we have created:

```
$ head -n 5 miR-564.fasta
hsa_miR.fasta:>hsa-miR-564 MIMAT0003228
hsa_miR.fasta-AGGCACGGUGUCAGCAGGC
--
ppy_miR.fasta:>ppy-miR-564 MIMAT0016009
ppy_miR.fasta-AGGCACGGUGGCAGCAGGC
```

We can see that the output of grep is the name of the original file where a match was found, followed by the line that contained the match. The -A1 option of grep also returned the line after the match (i.e., the sequence).

Knowing how to perform such simple loops using the Bash shell is very beneficial. However, Bash has a rather idiosyncratic syntax that does not lend itself well to performing more complex programming tasks. We will therefore cover general programming comprehensively in chapter 3, which introduces a programming language with a friendlier syntax, Python.

1.9 Tips, Tricks, and Going beyond the Basics

1.9.1 Setting a PATH in .bash_profile

Have you come across the error message command not found? You may have simply mistyped a command, or tried to invoke a program that is not installed on your machine. Maybe, however, your computer doesn't know the location of a program, in which case this can be resolved by adding the path (location) of a program to the PATH variable. Your computer uses $PATH to search for corresponding executable files when you invoke a command in the terminal. To inspect your path variable, type

```
# print path variable to screen
$ echo $PATH
```

You can append a directory name (i.e., location of a program) to your PATH by editing your .bash_profile. This file customizes your Bash shell (e.g., sets the colors of your terminal or changes the command-line prompt). If this hidden file does not exist in your home directory (check with ls -a), you can simply create it. Here is how to append to your computer's PATH variable ($PATH):

```
# add path to a program to computer's PATH variable
$ export PATH=$PATH:[PATH TO PROGRAM]
```

You can use which to identify the path to a program:

```
# identify the path to the grep command
$ which grep
/usr/bin/grep
```

Note that the order of elements in the PATH matters. If you have several versions of a program installed on your machine, the one that is found first (i.e., its location is represented earlier in the PATH) will be invoked.

1.9.2 Line Terminators

In text files, line terminators are represented by *nonprinting characters*. These are special characters that indicate white space or special formatting (e.g., space, tab, line break, nonbreaking hyphen). Unless you explicitly ask your editor to display them, they will not print to the screen (hence nonprinting). Unfortunately, different platforms use different symbols to indicate line breaks. While Unix-like systems use a line feed (\n), Windows uses a carriage return and linefeed combination (\r\n).

Many text editors autodetect the line terminator and display your file correctly (i.e., alter your file). However, if you encounter text that looks like a single long line, or displays ^M where a line break should be, you might want to change the nonprinting symbol for the line terminator.

Also, be careful when copying text between files with different line terminators—you might end up with a hybrid that will be difficult to deal with later on. When working with an unknown text file, use file to determine the type of line terminator before opening the file.

1.9.3 Miscellaneous Commands

Without much ado we want to provide some pointers to interesting topics and commands that might come in handy. Please refer to the documentation for usage and examples. Note that some commands are not available on all platforms, but can be installed.

history	List the last commands you executed.[13]
time [COMMAND]	Time the execution of a command.
wget [URL]	Download the web page at [URL].[14]
open	Open file or directory with default program; use xdg-open in Ubuntu or start in Windows Git Bash.
rsync	Synchronize files locally or remotely.
tar and zip	(Un)compress and package files and directories.
awk and sed	Powerful command-line text editors for much more complex text manipulation than tr.

---

13. In Git Bash all commands are listed.

14. Available in Ubuntu; for OS X look at curl, or install wget (see computingskillsforbiologists.com/wget).

xargs   Pass a list of arguments to other commands; for example, create a file for each line in files.txt:

`cat files.txt | xargs touch`

1.10 Exercises

1.10.1 Next Generation Sequencing Data

In this exercise we work with next generation sequencing (NGS) data. Unix is excellent at manipulating the huge FASTA files that are generated in NGS experiments.

FASTA files contain sequence data in text format. Each sequence segment is preceded by a single-line description. The first character of the description line is a "greater than" sign (>).[15]

The NGS data set we will be working with was published by Marra and DeWoody (2014), who investigated the immunogenetic repertoire of rodents. You will find the sequence file Marra2014_data.fasta in the directory CSB/unix/data. The file contains sequence segments (contigs) of variable size. The description of each contig provides its length, the number of reads that contributed to the contig, its isogroup (representing the collection of alternative splice products of a possible gene), and the isotig status.

1. Change directory to CSB/unix/sandbox.
2. What is the size of the file Marra2014_data.fasta?[16]
3. Create a copy of Marra2014_data.fasta in the sandbox and name it my_file.fasta.
4. How many contigs are classified as isogroup00036?
5. Replace the original "two-spaces" delimiter with a comma.
6. How many unique isogroups are in the file?
7. Which contig has the highest number of reads (numreads)? How many reads does it have?

1.10.2 Hormone Levels in Baboons

Gesquiere et al. (2011) studied hormone levels in the blood of baboons. Every individual was sampled several times.

---

15. See computingskillsforbiologists.com/fasta for more details on the FASTA file format.

16. Note that the original sequence file is much larger! We truncated the file to 1% of its original size to facilitate the download.

1. How many times were the levels of individuals 3 and 27 recorded?
2. Write a script taking as input the file name and the ID of the individual, and returning the number of records for that ID.
3. [**Advanced**][17] Write a script that returns the number of times each individual was sampled.

1.10.3 Plant–Pollinator Networks

Saavedra and Stouffer (2013) studied several plant–pollinator networks. These can be represented as rectangular matrices where the rows are pollinators, the columns plants, a 0 indicates the absence and 1 the presence of an interaction between the plant and the pollinator.

The data of Saavedra and Stouffer (2013) can be found in the directory CSB/unix/data/Saavedra2013.

1. Write a script that takes one of these files and determines the number of rows (pollinators) and columns (plants). Note that columns are separated by spaces and that there is a space at the end of each line. Your script should return

   ```
   $ bash netsize.sh ../data/Saavedra2013/n1.txt
   Filename: ../data/Saavedra2013/n1.txt
   Number of rows: 97
   Number of columns: 80
   ```

2. [**Advanced**][18] Write a script that prints the numbers of rows and columns for each network:

   ```
   $ bash netsize_all.sh
   ../data/Saavedra2013/n10.txt 14 20
   ../data/Saavedra2013/n11.txt 270 91
   ../data/Saavedra2013/n12.txt 7 72
   ../data/Saavedra2013/n13.txt 61 17
   ...
   ```

---

17. This task requires being able to capture the output of a command within a script (see, e.g., computingskillsforbiologists.com/captureoutput) and writing a "loop" iterating through all IDs (see, e.g., computingskillsforbiologists.com/bashloops).

18. This exercise requires writing a loop within a script.

3. Which file has the largest number of rows? Which has the largest number of columns?

1.10.4 Data Explorer

Buzzard et al. (2016) collected data on the growth of a forest in Costa Rica. In the file Buzzard2015_data.csv you will find a subset of their data, including taxonomic information, abundance, and biomass of trees.

1. Write a script that, for a given CSV file and column number, prints
   - the corresponding column name;
   - the number of distinct values in the column;
   - the minimum value;
   - the maximum value.

   For example, running the script with

   ```
   $ bash explore.sh ../data/Buzzard2015_data.csv 7
   ```

   should return

   ```
   Column name:
   biomass
   Number of distinct values:
   285
   Minimum value:
   1.048466198
   Maximum value:
   14897.29471
   ```

1.11 References and Reading

Books

J. Peek et al., *Unix Power Tools*, O'Reilly Media, 2005.
Tips & tricks to help you think creatively about Unix and solve your own problems.

A. Robbins & N. H. F. Beebe, *Shell Scripting*, O'Reilly Media, 2002.
Learn how to combine the fundamental Unix text and file processing commands to crunch data and automate repetitive tasks.

Online Resources

There are many good tutorials on Unix and shell scripting. Azalee Bos of Software Carpentry has a nice video tutorial on YouTube (shell starts at minute 11):
computingskillsforbiologists.com/unixvideo.

Also check out the material provided by Software Carpentry:
computingskillsforbiologists.com/unixtutorials.

The website SHELLdorado contains a list of useful shell scripts, as well as a series of tips & tricks:
shelldorado.com.

There is a series of tutorials on basic Unix commands:
computingskillsforbiologists.com/shelltutorials.

Many Unix commands are listed on Wikipedia:
computingskillsforbiologists.com/unixcommands.

## CHAPTER 2

Version Control

2.1 What Is Version Control?

A version control system (VCS) is a tool to record and organize changes to a set of files and directories (e.g., the directory containing one of your projects). Over time, a version control system builds a database storing all the changes you perform (a *repository*), making the whole history of the project available.

When you start working on a new project, you tell your VCS to keep track of all the changes, additions, and deletions. At any point, you can *commit* the changes, effectively building a snapshot of the project in the repository. This snapshot of the project is then accessible—you can recover previously committed versions of files, including metadata such as who changed what, when, and why. Also, you can easily start tracking files for an existing project.

Version control is especially important for collaborative projects: everybody can simultaneously work on the project—and even on the same file. Conflicting changes are reported, and can be managed using side-by-side comparisons. The possibility of *branching* allows you to experiment with changes (e.g., Shall we rewrite the introduction of this paper?), and then decide whether to *merge* them into the project.

2.2 Why Use Version Control?

Version control is fundamental to keeping a project tidily organized and central to making scientific computing as automated, reproducible, and easy to read and share as possible.

Many scientists keep backup versions of the same project over time or append a date/initials to different versions of the same file (e.g., various drafts of the same manuscript). This "manual" approach quickly becomes unmanageable, with simply too many files and versions to keep track of in a timely and organized manner. Version control allows you to access all previously committed versions of the files and directories of your project. This means that it is quite easy to undo short-term changes: Having a bad day?—scrap all modifications and go back to yesterday's version! You can also access previous stages of the project (I need to access my manuscript and all the associated files in exactly the state they were when I sent my manuscript for review three months ago). Going back to previously saved versions is easy with a VCS, but much more difficult with Dropbox or Google Drive.

Version control is useful for small, and essential for large, collaborative projects. It vastly improves the workflow, efficiency, and reproducibility. Without it, it is quite easy to lose track of the status of a manuscript (Who has the most recent version?), or lose time (I cannot proceed with my changes, because I need to wait for the edits of my collaborators).

Version control might look like overkill at first. However, with a little bit of practice you will get into the habit of running a few commands before you start working on a project, and again once you are done working on it—a small price to pay, considering the advantages. Simply put, using version control makes you a more organized and efficient scientist. Our laboratory adopted version control for all projects in 2010, and sometimes we wonder how we managed without.

Throughout this introduction, we illustrate the advantages of using version control for conducting scientific research, and assume that each repository is a scientific project. At first, we work with *local* repositories, meaning that all the files are stored exclusively on your computer. Then, we introduce *remote* repositories, which are also hosted on a web server, making it easy for you to share your projects with others (or work on the same project from different computers).

2.3 Getting Started with Git

For this introduction to version control we use Git, which is one of the most popular version control systems. Git is also free software, and is available for all computer architectures. Many good tutorials are available online, and many websites will host remote Git repositories for free.

Other options you might want to consider are Mercurial (very similar to Git) and Subversion version control (svn), which is an older system—but still widespread.

There are two main paradigms for VCSs allowing multiple users to collaborate: in a *centralized* VCS (e.g., svn), the whole history of a project is stored exclusively on a server, and users download the most current snapshot of the project, modify it, and send the changes to the server; in a *distributed* VCS (e.g., Git), the complete history of the repository is stored on each user's computer.

Git was initially developed by Linus Torvalds (the "Linu" in Linux), exactly for the development of the Linux kernel. It was first released in 2005 and has since become the most widely adopted version control system.

2.3.1 Installing Git

You can find architecture-specific instructions in the directory CSB/git/ installation. The instructions also contain the commands you should run to configure the system.

2.3.2 Configuring Git after Installation

The first time you use Git (or whenever you install Git on a new computer), you need to set up the environment. To store your preferred user name and email, open a terminal and type

```
$ git config --global user.name "Charles Darwin"
$ git config --global user.email crdarwin@royalsociety.org
```

Optionally, you can set up a preferred text editor (e.g., gedit or emacs), which will be used to write the messages associated with your commits:

```
$ git config --global core.editor gedit
```

You can activate colors in your terminal to make it easier to see the changes to your files:

```
$ git config --global color.ui true
```

To check all of your settings and see all available options, type

```
$ git config --list
```

2.3.3 How to Get Help in Git

For a brief overview of common Git commands, open your terminal and type

```
$ git help
```

Funny fact: if you type

```
$ man git
```

you will see that the name of the program is git - the stupid content tracker. Undoubtedly, Git behaves stupidly in the sense that the system tracks any content, without being selective (which can be a good or a bad thing). The manual page contains a description of all the commands, but it is much more pleasing to read them online.[1]

2.4 Everyday Git

To illustrate the basic operations in Git, we consider the case of a local repository: all the versions are stored only in your computer, and we assume that you are the only person working on the project. Once familiarized with the basics of Git, we introduce the use of remote repositories for collaborative projects.

2.4.1 Workflow

Understanding how Git works means internalizing its workflow. Figure 2.1 depicts the typical day of a Git user.

As you can see, there are only a few commands that you need to master for everyday work. Let's try our hands at this workflow and create a simple repository. This is just for practice, so we create it in the CSB/git/sandbox directory:

```
$ cd ~/CSB/git/sandbox
$ mkdir originspecies
$ cd originspecies
```

---

1. A freely available book can be found at git-scm.com/book.

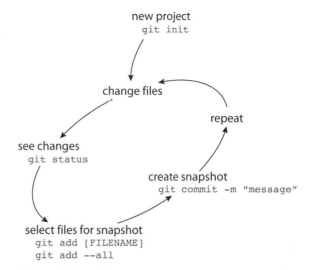

Figure 2.1. Basic Git workflow. When starting a new project, create a directory using terminal and initialize a repository. Start working on your files until you reach a milestone, or until you are done for the day. Check what has changed since your last snapshot. Decide which files to add to the new snapshot of the project (select particular files, or add everything). Create the snapshot by committing your changes, including a detailed description. Start changing the files again, add them to the new snapshot, and commit.

```
$ git init
Initialized empty Git repository in /tmp/originspecies/
    ↳ .git
```

We have moved to a newly created directory (originspecies), and run the command git init, which initializes an empty repository. You can set up a Git repository for an existing project by changing to the directory containing the project and typing git init in the terminal. We recommend always running pwd before initializing a Git repository, to confirm that you are in the correct directory. The last thing you want is to track changes to your entire computer because you happened to be in your root directory.

We can check the status of the repository by running

```
$ git status
On branch master

Initial commit
```

```
nothing to commit (create/copy files and use "git add" to
    ↳ track)
```

Now we create our first file:

```
$ touch origin.txt
```

and then start editing it. We could use a text editor, but for the moment let's stick to the command line (see section 1.5 for a review of basic Unix commands):

```
$ echo "An Abstract of an Essay on the Origin of Species
    ↳ and Varieties Through Natural Selection" > origin.
    ↳ txt
```

and check that our command went through:

```
$ cat origin.txt
An Abstract of an Essay on the Origin of Species and
    ↳ Varieties Through Natural Selection
```

Git does not track any file unless you tell it to do so. We can set the file origin.txt for tracking with the command

```
$ git add origin.txt
```

We can see that the status of the repository has changed:

```
$ git status
On branch master

Initial commit
```

```
Changes to be committed:
  (use "git rm --cached <file>..." to unstage)

    new file:   origin.txt
```

Every time we want to signal Git that a file needs to be tracked, we need to add it to the index of files in the repository using the command `git add [FILENAME]`. If we want simply to add all the files in a directory and subdirectories, use `git add .` (where the `.` means current directory). To add all the files contained in the repository directory (including "hidden" files), use `git add --all`.

Note that Git and similar systems are ideally suited to working with text files: `.csv` and `.txt` files, code (`.R`, `.py`, etc.), LaTeX manuscripts (`.tex`, `.bib`), etc. If you track *binary* files, Git will simply save a new version any time you change it, but you will not be able to automatically see the differences between different versions of the files.[2]

A text file can be read by a human when opened in any text editor (for instance, `less`, gedit, or emacs). A *binary* file is computer readable, but not human readable. Try opening a binary file (such as `.docx`, `.xls`, or `.pdf`) using the `less` command in Unix, or any text editor: you will see a lot of gibberish. Opening a binary file requires dedicated software that can interpret the binary code and display human-readable output.

Once we are finished creating, deleting, and modifying our files, we can create a snapshot of the project by *committing* the changes. When should one commit? Git's motto is "commit early, commit often." Every time you commit your changes, these are permanently saved in the repository. Ideally, every commit should represent a meaningful step on the path to completing the project (examples: "drafted introduction," "implemented simulation," "rewritten hill-climber," "added the references," …). As a rule of thumb, you should commit every time you can explain the meaning of your changes in a few lines.

Now it's time to perform our first commit:

---

2. Unless you really want to, in which case there are work-arounds: computingskillsfor biologists.com/customgit.

```
$ git commit -m "started the book"
[master (root-commit) 551b3bb] started the book
 1 file changed, 1 insertion(+)
 create mode 100644 origin.txt
```

where -m stands for "message." If no message is entered, Git will open your default text editor, so that you can write a longer message detailing your changes. These messages are very important! In fact, you can use them to navigate and understand the history of your project. Make sure you always spend a few extra seconds to document what you did by writing a meaningful and detailed message. In a few months, you will have forgotten all the details associated with a commit, so that your future self will be grateful for detailed messages specifying how and why your project has changed.

The history of the repository can be accessed by typing

```
$ git log
commit 6cadf65eff4cb8e2bb3ed8696025ceb2e33dff9a
Author: Charles Darwin <crdarwin@royalsociety.org>
Date:   Tue Jul 20 15:26:19 1858 -0000

    started the book
```

The long line of numbers and letters after the word commit is the "checksum" associated with the commit (i.e., a number Git uses to make sure that all changes have been successfully stored). You can think of this number as the "fingerprint" of the commit. When you follow along in your own terminal, naturally all checksums and dates will differ from what shown in the code boxes.

Now, let's change a tracked file:

```
$ echo "On the Origin of Species, by Means of Natural
    ↳ Selection, or the Preservation of Favoured Races in
    ↳ the Struggle for Life" > origin.txt
```

which is a much more powerful, albeit rather long, title. The repository has changed and we can investigate the changes using status:

```
$ git status
On branch master
Changes not staged for commit:
  (use "git add <file>..." to update what will be committed)
  (use "git checkout -- <file>..." to discard changes in
      ↳ working directory)

   modified:    origin.txt
```

This shows that (a) a file that is being tracked has changed, and (b) that these changes have not been *staged* (i.e., marked) to be committed yet. You can keep modifying the file, and once you are satisfied with it, use git add to make these changes part of the next commit. For example,

```
$ git add .
$ git commit -m "Changed the title as suggested by Murray"
```

You now see the new commit in the history of the repository:

```
$ git log
"commit" 6cadf65eff4cb8e2bb3ed8696025ceb2e33dff9a
Author: Charles Darwin <crdarwin@royalsociety.org>
Date:   Fri Aug 13 10:23:49 1858 -0000

    Changed the title as suggested by Murray

commit 5e0864724f2e64848a5b0d0364c1292610ceab61
Author: Charles Darwin <crdarwin@royalsociety.org>
Date:   Tue Jul 20 15:26:19 1858 -0000

    started the book
```

That's it! For 99% of your time working with Git, all you need to do is to follow these simple steps:

```
# when you are creating a new project
$ mkdir newproject
$ cd newproject
```

```
$ git init # initialize repository

# daily routine:
# (1) change the files
# (2) check status
$ git status
# (3) add files to the snapshot
$ git add --all
# (4) commit the snapshot
$ git commit -m "my descriptive message"
```

Besides these basic commands, Git allows you to do much more. We explore some of the more advanced features in the next sections.

Intermezzo 2.1

(a) Create the file todo.txt containing the line June 18, 1858: read essay from Wallace.
(b) Add the file to the next snapshot.
(c) Commit the snapshot, with message Added to-do list.

2.4.2 Showing Changes

If you want to see all the changes you made since the last commit, you can run

```
$ git diff
```

For example, suppose we edited the file origin.txt, and we want to see the changes we performed:

```
$ git diff
diff --git a/origin.txt b/origin.txt
index 2e6e57c..ff025c0 100644
--- a/origin.txt
+++ b/origin.txt
@@ -1 +1,11 @@
```

```
-On the Origin of Species, by Means of Natural Selection,
    ↳ or the Preservation of Favoured Races in the
    ↳ Struggle for Life
+On the Origin of Species,
+
+by Means of Natural Selection, or the Preservation of
    ↳ Favoured Races in the Struggle for Life
+
+BY CHARLES DARWIN, M.A.,
+
+FELLOW OF THE ROYAL, GEOLOGICAL, LINNAEAN, ETC.,
    ↳ SOCIETIES;
+AUTHOR OF 'JOURNAL OF RESEARCHES DURING H.M.S. BEAGLE'S
    ↳ VOYAGE ROUND THE WORLD.'
```

For your convenience, Git shows the changes between the previous version (marked a/origin.txt), and the current version (marked b/origin.txt). All additions to the files are marked by "+" and deletions by "−" at the beginning of each line. If you set up Git to use colors (git config color.ui true), the differences will be colored in red (deletions) and green (additions).

2.4.3 Ignoring Files and Directories

Often, there are specific types of files you want to ignore (e.g., temporary files, binary files obtained by compiling code that is already present in the repository, databases that are too large to be stored with the project).

You can tell Git to ignore certain types of files, or specific files and directories, by creating a file called .gitignore in the main directory of the repository (you can create/edit the file using any text editor). For example, here's a typical .gitignore file:

```
$ cat .gitignore
*~
*.tmp
binaries/
largedataset.db
```

which means that Git should ignore every file whose name ends with "tilde" (~, temporary files), has extension .tmp, is contained in the directory binaries, or is called largedataset.db.

2.4.4 Moving and Removing Files

When you want to move or remove files or directories under version control, you should let Git know so that it can update the index of files to be tracked. Doing so is quite easy: simply put git in front of the command you would run to perform the operation in Unix:

```
$ git rm filetorem.txt
$ git rm *.txt
$ git mv myoldname.txt mynewname.csv
```

2.4.5 Troubleshooting Git

At first, your Git workflow may not be as linear and straightforward as in the examples above. Conflicts or mistakes occur, and here is a brief guide on how to deal with them.

Amending an Incomplete Commit

The most common mistake is to commit something only to find out that your snapshot is not fully functional (e.g., you forgot to add a line to your code or to uncomment a part of the manuscript), or that you forgot to add (or remove) this or that file. You could create another commit (e.g., git commit -m "fixed a few bugs"), but in the long run these small commits are annoying, as they break up the logical history of your repository. You can "add what you forgot" to a commit by running

```
$ git add forgottenfile.txt
$ git add fixedcode.py
$ git commit --amend
```

which, as the name implies, amends your previous commit, resulting in a single commit instead of two. Many Git purists abhor the use of amend, as it "alters history" and deletes the previous commit: we report the command here, because sometimes it is the most practical thing to do.

Unstaging Files

The second most common mistake is to add a modification of a file to the repository that is not intended to be part of the commit (e.g., some file you modified only for debugging a certain problem).

In Git, each file can be in one of three states:

Modified   You have modified the file, but not marked it to be committed yet.

Staged   You have marked the changes to the file to be added to the next snapshot, using the command git add. You can think of the staging area as the "loading dock": the changes are ready to be shipped, but haven't been shipped yet.

Committed   The file is stored in your repository, as part of a snapshot. Snapshots are saved in a special directory (the .git directory within your project directory).

Did you notice that Git tries to help you to type reasonable commands? When you type git status it not only tells you the status of each file but will often also suggest how to proceed. For instance, you accidentally staged a file by typing git add ., but now you would like to "unstage" it. Type git status and Git will provide a helpful suggestion:

```
(use "git reset HEAD <file>..." to unstage)
```

Aha! You can use the command reset to remove the modifications to the file from the staging area. If you now follow Git's suggestion and type

```
$ git reset HEAD filetounstage.py
```

then the file will revert to the state it was in during the last commit. HEAD is a reference to the commit that we are currently working with.

Deleting Changes and Reverting to the Last Commit

Sometimes, changes can go horribly wrong, such that you would like to scrap everything and abandon all the changes you made to one (or more) file(s). If you haven't staged these changes, this can be accomplished simply by typing

```
$ git checkout filetoreset.txt
```

This command fetches a "fresh" copy from the repository, meaning that the file will be in the version that was last committed. Use this command with caution, as the changes you have been making are irrevocably lost.

2.5 Remote Repositories

So far, we have been working with local repositories, hosted in only one computer. When collaborating (or when working on the same project from different computers), it is convenient to host the repository on a server, so that collaborators can see your changes, and add their own.

There are two main options: (a) set up your own server (this is beyond the scope of this introduction, but see section 2.8 for pointers on how to do this); (b) host your repositories on a website offering this service (for a fee, or for free).

The most popular option for hosting open-source projects (i.e., where the whole world can see what and when you commit) is GitHub (github.com), which will store all your public repositories for free. Private repositories (i.e., repositories whose access is restricted to authorized users) are available for a fee. Both GitHub (education.github.com) and Bitbucket (bitbucket.org) offer free private repositories for academic users. Besides hosting the repositories, these and similar services help with setting up your repositories, and include extra features such as a wiki and issue tracker for your projects. Using an online hosting service also makes visualization of the changes and the history of the project very convenient, thanks to a browser-based graphical interface.

If you're the main developer of the project, you first need to set up your repository using one of the services mentioned above. To download a copy of your repository, your collaborators need to clone the repository.

You already performed this operation when you downloaded the repository associated with this book in the initial setup (see computingskillsfor biologists.com/setup):

```
# choose where to download the new repository
$ cd ~
# download a local copy of a repository
$ git clone https://github.com/CSB-book/CSB.git
```

Once the repository is in place, you need to add only two new commands to your Git workflow: pull and push. When you want to work on a project that is tracked by a remote repository, you *pull* the most recent version from the server and work on your local copy using the commands illustrated in figure 2.1. When you are done, you *push* your commits to the server so that other users can see them. The complete workflow will look like this:

```
# get the latest version of the project
$ git pull

# start making changes; when done run
$ git add --all
# time to commit:
$ git commit -m "A meaningful message"
# rinse and repeat...

# ok, everyone should see my changes
$ git push
```

Note that the interaction with the server happens only when you pull and push. In between these commands, you are in control, and can decide when and what to communicate to your collaborators.

The merge with the remote repository aborts if there are conflicts with unstaged versions in your local repository. You have two options to clean your work area before pulling from the remote repository: you can either stage and commit your changes as usual, or *stash* your changes. The latter is useful when you need to pull from the remote repository (e.g., to look at the changes of your collaborator) but you do not want to commit the incomplete task you were just working on. As the name implies, you are stashing these incomplete changes away, which leaves you with a clean working directory (i.e., everything is in the state of the last commit). You can then pull from the remote repository and *apply* your stashed changes back to the files. Here are the commands:

```
# stash changes to tracked files
$ git stash
# add option -u to include untracked files
# check your status
$ git status
# pull from the remote repository
$ git pull
# apply stashed changes
$ git stash apply
```

2.6 Branching and Merging

Most VCSs allow you to *branch* from the main development of the project (i.e., experiment freely with your project without messing up the original version). It is like saying "Save As" with a different name—but with the possibility of easily merging the two parallel versions of the project. Typical examples of a branching point in scientific projects are (a) you want to try a different angle for the introduction of your manuscript, but you are not sure it's going to be better; (b) you want to try to rewrite a piece of code to see whether it will be faster, but you certainly do not want to ruin the current version that is working just fine; (c) you want to keep working on the figures while your collaborator is editing the manuscript. In these cases, you are working on an "experimental feature," while leaving the main project unaltered (or while other people are working on it). Once you are satisfied with your changes, you would like to *merge* them with the main version of the project.

To explore branching, we create a new repository in CSB/git/sandbox. Instead of typing all commands to set up the repository, you can take a short-cut by executing the script[3] create_repository.sh which is located in the data directory.

We first look at the script to see what it does:[4]

---

3. Shell scripts are introduced in section 1.7.

4. For security reasons, it is always a good idea to confirm that a shell script does what you expect it to do (e.g., does not install malicious software).

```
$ less ../data/create_repository.sh
# hit q to exit the file pager
```

Execute the script to set up the repository:

```
# execute script located in CSB/git/data
$ ./../data/create_repository.sh
```

The shell script creates the directory branching_example, initiates a Git repository, and creates two files that are committed individually. Let's have a look at the new repository that you now have in your sandbox:

```
$ cd branching_example
$ git log --oneline --decorate
bc4831a (HEAD -> master) Drafted paper
7b1a79e Code ready
# your checksums will differ
```

You can see that there are two commits: Code ready, and the more recent commit Drafted paper at the top of the list. Right now, the main "trunk" of the project (i.e., master) is associated with the second commit (Drafted paper). The HEAD pointer tells us that Drafted paper is the current commit on the master branch.

Figure 2.2 provides a schematic drawing of the repository branching_example, including the commits and branch fastercode that we are about to construct.

Now, suppose that you want to try amending the code (e.g., to make it run faster), but don't want to touch the code that is already working. You need to *branch*:

```
# create a new branch
$ git branch fastercode
# check log of repository
$ git log --oneline --decorate
bc4831a (HEAD -> master, fastercode) Drafted paper
7b1a79e Code ready
```

We have created the new branch fastercode. When naming branches, use either single words or connect words with underscores to avoid spaces in the

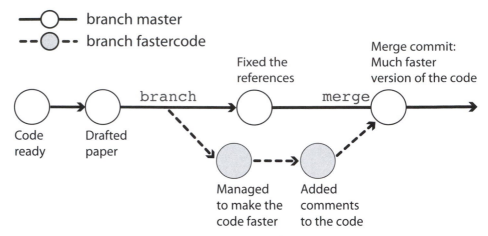

Figure 2.2. Schema of *branching* in Git. After two commits to the master branch, the branch fastercode is created and receives two additional commits without affecting the master branch. Meanwhile master receives one additional commit. Eventually, branch fastercode is *merged* into master and can be deleted. After the merge, the master branch contains all edits that were committed to either branch.

name of the branch. Calling the log of our repository after creating a new branch shows that master, fastercode, and HEAD are all associated with the same commit. Typing

```
# list branches in your repository
$ git branch
  fastercode
* master
```

shows that you are currently working on master (hence, the *). To switch to the new branch, type

```
# switch branch
$ git checkout fastercode
Switched to branch 'fastercode'
# list branch
$ git branch
* fastercode
  master
```

which means that now you can experiment freely. Suppose that you edit the code:

```
$ echo "Yeah, faster code" >> code.txt
```

and commit the changes:

```
# add to staging
$ git add code.txt
# commit changes
$ git commit -m "Managed to make code faster"
[fastercode 56d1830] Managed to make code faster
 1 file changed, 1 insertion(+)
```

Running the log,

```
$ git log --oneline --decorate
56d1830 (HEAD -> fastercode) Managed to make code faster
bc4831a (master) Drafted paper
7b1a79e Code ready
```

shows that you are working on the fastercode branch, and that you have a new commit that is not part of the master branch. Right at this moment, your advisor tells you that you should add a bunch of references to the paper—while you were just about to make some more changes to the faster code. No problem—simply switch to the master branch, fix the references and commit:

```
$ git checkout master
# fix the references
$ echo "Marra et al. 2014" > references.txt
# add to staging and commit
$ git add references.txt
$ git commit -m "Fixed the references"
```

How are we doing? You can ask Git to print you a graph summarizing the state of all branches:

```
# show log of repository as graph
$ git log --oneline --decorate --all --graph
* 7faa62d (HEAD -> master) Fixed the references
| * 56d1830 (fastercode) Managed to make code faster
```

```
|/
* bc4831a Drafted paper
* 7b1a79e Code ready
```

The | and / represent the branches and * the commits. You can see that you are currently working on the master branch (hence, the HEAD), and that you have a commit that is in master, but not in fastercode (Fixed the references), and one that is in fastercode, but not in master (Managed to make code faster). Having fixed the references, we can switch back to fastercode, and keep working on it (e.g., writing good comments for the code).

```
$ git checkout fastercode
# edit the code
$ echo "# My documentation" >> code.txt
# add to staging and commit
$ git add code.txt
$ git commit -m "Added comments to the code"
[fastercode 7237491] Added comments to the code
 1 file changed, 1 insertion(+)
```

Now, let's see how the graph is looking:

```
$ git log --oneline --decorate --all --graph
* 7237491 (HEAD -> fastercode) Added comments to the code
* 56d1830 Managed to make code faster
| * 7faa62d (master) Fixed the references
|/
* bc4831a Drafted paper
* 7b1a79e Code ready
```

which shows that now fastercode (where we are working) has two commits that are not part of master. If you are satisfied with the changes, you might want to merge. Think of *merging* as reaching out to another branch, grabbing files from it, and merging it into the branch you are currently on. So the first thing to do is switch to the master branch:

```
# switch to master branch
$ git checkout master
Switched to branch 'master'
```

```
# merge code from branch fastercode
$ git merge fastercode -m "Much faster version of code"
Merge made by the 'recursive' strategy.
```

In this case, the merge happened automatically, as there were no conflicts (i.e., we modified different files in the two branches). Here's the graph after the merge:

```
$ git log --oneline --decorate --all --graph
*   bf42170 (HEAD -> master) Much faster version of code
|\
| * 7237491 (fastercode) Added comments to the code
| * 56d1830 Managed to make code faster
* | 7faa62d Fixed the references
|/
* bc4831a Drafted paper
* 7b1a79e Code ready
```

We merged the branch, and no longer need it. We delete it by typing

```
$ git branch -d fastercode
Deleted branch fastercode (was 7237491).
$ git log --oneline --decorate --all --graph
*   bf42170 (HEAD -> master) Much faster version of code
|\
| * 7237491 Added comments to the code
| * 56d1830 Managed to make code faster
* | 7faa62d Fixed the references
|/
* bc4831a Drafted paper
* 7b1a79e Code ready
```

To summarize, here's how you work with branches:

```
# create a new branch
$ git branch mybranch
# see that you are in master
$ git branch
```

```
# move to new branch
$ git checkout mybranch

# start working on the branch
# ...
# add and commit changes
$ git add --all
$ git commit -m "my message"

# once you are satisfied, and want to include
# the changes into the main project,
# move back to main trunk
$ git checkout master
# then merge the branch
$ git merge mybranch -m "message for merge"
# and [optionally] delete the branch
$ git -d mybranch
```

Viewing Past Revisions

You can go back to see what your project looked like in the past by *checking out* an old commit. For example, take the project we have been working on:

```
$ git log --oneline --decorate --all --graph
*   bf42170 (HEAD -> master) Much faster version of code
|\
| * 7237491 Added comments to the code
| * 56d1830 Managed to make code faster
* | 7faa62d Fixed the references
|/
* bc4831a Drafted paper
* 7b1a79e Code ready
```

and suppose we want to go back in time to the commit Drafted paper. To do so, simply run

```
# remember to use correct checksum
$ git checkout bc4831a
```

where bc4831a is the initial part of the checksum for the commit, as displayed by git log. You can look around, run code, compile files, etc. Nothing you are doing is saved in the repository. Once you are done looking around, you can go back to the future by typing

```
$ git checkout master
```

Note that if you change anything, this will automatically create a branch, as you cannot modify the past!

Resolving Merge Conflicts

In most cases, Git will merge branches to the main project development without any issue. However, conflicts can arise if you and your collaborators changed the exact same line of a file in different ways (in the main trunk, and in the branch you want to merge). In these cases, Git does not know which changes should be retained, and which should be foregone. Thus, it will abort the merge and signal the problem, asking you to resolve the conflict.

```
$ git merge -m "We need to merge"
Auto-merging code.txt
CONFLICT (content): Merge conflict in code.txt
Automatic merge failed; fix conflicts and then commit the
    ↳ result.
```

You can type git status to display a detailed description of the conflict. To resolve it, open the file containing the conflicting edits in a text editor. Git automatically inserts markers around the lines that are in conflict. The area that needs to be fixed begins with <<<<<<< and ends with >>>>>>>. In between these markers are both versions of the conflicting lines, separated by =======. Your task is to delete these markers, and resolve the conflict by keeping either version of the lines (or write an entirely new one). When you have finished fixing all merge conflicts, stage and commit the file with a new commit message. This resolves the merging conflict and you can go on with your work.

Intermezzo 2.2

(a) Move to the directory CSB/git/sandbox.

(b) Create a thesis directory and turn it into a Git repository.

(c) Create the file introduction.txt with the line "The best introduction ever."

(d) Add the file introduction.txt to the next snapshot and commit with the message "Started introduction."

(e) Create the branch newintro and change into it.

(f) Change the contents of introduction.txt, create a new file with the name methods.txt, and commit.

(g) Look at the commit history of the branches.

(h) Change to the branch master, merge, and confirm that the changes you performed within the branch newintro are now also part of the branch master.

(i) Delete the branch newintro.

2.7 Contributing to Public Repositories

Much software for science is found in public repositories, typically hosted by web-based hosting services such as GitHub or Bitbucket. A public repository allows anyone to contribute to the stored project—though what gets included is decided by the developers who initially set up the repository.

Suppose that you want to contribute a new solution to the exercises in this book. You are not one of the administrators of the repository, and therefore what you need to do first is to create a *fork*, which is simply a copy of a repository, which you can then manage—you can experiment freely without affecting the original repository. In GitHub, creating a fork is very easy—just click on the "fork" button in the top-right corner of the landing page for the repository you want to contribute to. This will create a copy of the repository in your GitHub account. You can then manage this forked repository as if it were your own and, for example, clone it to your local machine (described in section 2.5).

Here we deal with the case in which you want to make a quick modification (e.g., fix a typo, add a file). If your modification is going to take a long time, you need to ensure that you are keeping up with the changes that are made in the original repository by keeping them synchronized.[5]

---

5. Read the documentation at computingskillsforbiologists.com/forkrepository.

Submitting changes to be incorporated into the original repository is called a *pull request*. The maintainers of the original repository can accept the changes (by merging them into their repository), or reject them.

To facilitate their work, you typically work with a branch, containing exclusively the changes you want to propose. Here's the workflow:

```
# clone the fork of the original repo to your computer
$ git clone https://github.com/YourName/CSB.git
$ cd CSB
# all branches should come from master
$ git checkout master
# create a new branch
$ git branch better_solution
$ git checkout better_solution
# make changes
# ...
# and when you're done, add and commit
$ git add --all
$ git commit -m "Provide a detailed description of your
    ↳ changes"
# push the new branch to your fork
$ git push origin better_solution
```

Now you're ready to submit your pull request. Go to the GitHub page of your fork, and click on "New pull request". In the box "compare:", choose your new branch. GitHub will let you know whether there are conflicts, or whether your branch could automatically be merged into the original project. Clicking on "Create pull request" will open a page in which you should describe the changes you've made, and why these improve the project. Clicking again on "Create pull request" will send the request to the maintainers for approval. Congratulations! You've just contributed to making science more open and transparent.

2.8 References and Reading

Books and Tutorials

There are very many good books and tutorials on Git. We are particularly fond of *Pro Git*, by Scott Chacon and Ben Straub. You can either buy a physical copy of the book or read it online for free at

git-scm.com/book.

Both GitHub and Atlassian (managing Bitbucket) have their own tutorials:
guides.github.com,
atlassian.com/git/tutorials.

A great way to try out Git in 15 minutes:
try.github.io.

Software Carpentry offers intensive on-site workshops and online tutorials:
swcarpentry.github.io/git-novice.

A visual Git reference:
computingskillsforbiologists.com/visualgit.

If you want to set up your own server for your Git repositories, have a look at computingskillsforbiologists.com/gitserver.

Graphical User Interface

There are several graphical user interfaces for Git:
git-scm.com/downloads/guis.

Scientific Articles

Many articles advocate the use of version control in biology. Among the recent ones, you can read the wonderful paper by Wilson et al. (2014) on implementing best practices in scientific computing, and the articles by Ram (2013) and Blischak et al. (2016), focusing on Git.

The British Ecological Society has produced a wonderful document on reproducibility of ecological data analysis (computingskillsforbiologists.com/reproduciblecode), with version control playing a key role.

CHAPTER 3

• • • • • • • • • • • •

Basic Programming

3.1 Why Programming?

There is almost no need to advocate good programming skills in the sciences—you probably picked up this book exactly because you want to become a more proficient programmer. Often, no software is available to perform exactly the analysis you have in mind. Knowing how to program can ease this problem, as you can write your own software that does precisely what is needed.

Writing your analysis in a program also makes your findings easier to reproduce, understand, and extend—especially compared to the case where the analysis is performed through a graphical user interface, with the user clicking through forms and windows to obtain the desired result. If you organize your data and code properly, and you automate the whole pipeline, anybody (anywhere) will be able to reproduce your findings exactly.

3.2 Choosing a Programming Language

Hundreds of programming languages are currently available. Which ones should a biologist choose? Note the use of the plural: as for spoken languages, knowing more than one language brings great advantages. In fact, each programming language has been developed with a certain audience and particular problems in mind, resulting in certain strengths and weaknesses. Acquiring a basic understanding of several languages allows you to choose the more appropriate language for each task. We do not recommend becoming a "jack of all programming languages and master of none," but at the same time using your favorite language for every problem might be counterproductive: though you can certainly find a way to solve the problem at hand, the solution might be inefficient or overly complex.

Most programming needs in the sciences can be solved with a combination of three programming languages:

1. A modern, easy-to-write language for data manipulation, prototyping, and programming tasks that do not entail several million operations. Common choices are Python, Perl, and Ruby. Two of the main advantages of these languages are that the code is platform independent, and you do not need to specify the type of a variable before using it (more on this later). The price to pay is typically in terms of speed of execution, which can be quite slow. Here, we cover the basics of Python because of its emphasis on readability, its flexibility, and its popularity.

2. Mathematical/statistical software with programming capabilities, like R, Mathematica, or MATLAB. A relatively new language in this arena is Julia, which aims to combine the simplicity of R with the speed of C. You can use this second language to perform mathematical and statistical analysis and data manipulation, and to draw figures for your publications. In chapters 8 and 9 we introduce R, which is very popular among biologists. There is also growing interest in R among people working in data science and data visualization (meaning that if academia is not for you, you can get a well-paid job thanks to your programming skills).

3. A language for the heavy lifting of your analysis (i.e., programs that require millions or billions of operations, such as simulations). Typically, these languages rely on a compiler, a special program that translates the source code into an executable, machine-specific binary program. These implementations tend to be very fast, especially for languages in which you have to specify the type of a variable before using it. A popular choice is C, as it provides fantastic libraries for scientific computing (e.g., the GNU Scientific Library). Other possibilities are FORTRAN or Go (a programming language launched by Google). If you need objects (e.g., you are writing an agent-based model, or building a graphical user interface), consider Objective-C, Java, Swift, or C++. In this introductory book, we are not covering any of these languages— but what you will learn in this and following chapters will surely make them easier to learn.

Though learning multiple languages might sound overwhelming, once you have mastered one language it is much easier to learn another one. Here we emphasize basic aspects that are common to most programming languages (e.g., dividing a complex problem into its basic constituents), and we stress good programming practices (debugging, profiling, unit testing), which are going to be useful for whatever programming language you might choose.

One might think that writing a faster program in C is always preferable to a slower program in Python. The website benchmarksgame.alioth.debian.org reports that a program simulating the n-body problem takes about 10 seconds to run in C, and 1000 seconds (about 17 minutes) in Python. Does this mean that everybody should just code in C? No, unless you need to repeat the same operations many, many times. When you choose a programming language to solve a computational problem, the time needed to run the program is not the only variable you should take into consideration. Rather, you should care about the total time elapsed before you obtain a result, which is the sum of the times spent programming, debugging, and running. In most cases, writing and debugging the code takes considerably more time than running it! This means that a "slower" programming language can be the better choice if the ease of the language allows you to write and debug your code faster.

3.3 Getting Started with Python

In this and following chapters, we introduce the basic syntax of Python along with examples that use biological data sets. We choose Python because (a) it is a carefully thought-out language, with great string manipulation capabilities (e.g., for bioinformatics), web interfaces (e.g., for downloading data from the internet), and useful data types such as sets and dictionaries; (b) it is freely available and well documented; and (c) it is easy to adopt a good programming style while writing Python code.

In this chapter, we introduce basic programming concepts, such as variable assignment, showcase the main data structures, and explain how to change the flow of a program; in chapter 4, we cover more advanced topics such as user-defined functions and tools that will make you write better code; in chapter 5, we deal with text mining using regular expressions; finally, in chapter 6 we focus on the development of scientific applications. Though all the examples are quite specific to Python, the main concepts are relevant to all programming languages.

3.3.1 Installing Python and Jupyter

Chances are you already have a copy of Python installed on your computer—even if you think you've never used it. However, we have compiled a platform-specific guide to installing Python in CSB/python/installation/

install.md in a way that will make following the rest of the material easier.[1] In particular, we use the package management system Anaconda, which greatly facilitates the installation and updating of Python packages.

For this introduction, we are going to use Jupyter "notebooks" that run Python in your web browser, so that the interface looks the same across operating systems. There are two ways to launch a new Jupyter notebook. Either click on the IPython notebook icon or open a terminal and type jupyter notebook. In both cases, Jupyter will start in your default internet browser, listing the files in the directory. Navigate to the location where you want a new notebook to be opened. Click the button "New" in the upper right-hand corner and select "Python 3" to open a new notebook.

In Jupyter, you type commands in the box next to In []:. To execute the command, press Shift+Enter. For clarity, in the text we start each command with In []:, but you don't have to type it—it is just for your reference. When commands span multiple lines, we omit the In []:. Lines starting with # are comments. To execute the code of one input cell, position your cursor in it and press Shift+Enter. If execution of that cell produces an output to the screen, it is shown in the next line, starting with an Out[]:.

You can save the notebook using the icons or the "File" menu, export the notebook as a PDF, or save it in different formats. For example, you might want to download the code contained in the notebook as a plain Python script, using "File" → "Download As" → Python (.py). In this way, the code can be run in machines that do not have Jupyter installed. One nice feature of these notebooks is that you can document your code using Markdown,[2] making your code easier to understand at a later point. To write a note in Markdown, just press Esc+M from within a cell.

3.3.2 How to Get Help in Python

We recommend the website docs.python.org as your first point of reference: it provides accurate, up-to-date, and extensive documentation. For each command, you will find a description in plain English, usage examples, and links to related projects. In addition, you can seek help on websites such

---

1. For the material in this book, we use version 3 of Python. Previous versions are not fully compatible—you can see a list of the main changes between versions 2 and 3 at computingskillsforbiologists.com/python2to3.

2. A simple markup language; see computingskillsforbiologists.com/markdownbasics for basic syntax.

as stackoverflow.com. Within your Jupyter notebook you can access a brief description by executing help("MY_COMMAND").

3.3.3 Simple Calculations with Basic Data Types

We can use Python as an oversized calculator. Create a new Jupyter notebook and type

```
In [1]: 2 + 2 # addition
Out[1]: 4
In [2]: 2 * 2 # multiplication
Out[2]: 4
In [3]: 3 / 2 # division
Out[3]: 1.5
In [4]: 3 // 2 # integer division
Out[4]: 1
In [5]: 2 > 3 # logical operators return a Boolean value
Out[5]: False
In [6]: 2 == 2 # equals
Out[6]: True
In [7]: 2 != 2 # is different
Out[7]: False
In [8]: "my string" # quotes define strings
Out[8]: 'my string'
```

As we just saw, Python stores values according to different built-in types, for example integers, floats (real numbers), Booleans (True, False), and strings. You can use either single or double quotation marks to define strings in Python (but make sure you use the same type of quotation marks to open and close the string). You may even use triple quotes if your string already contains single and double quotes—for example, when expressing lengths using feet and inches. Unlike other programming languages, Python automatically interprets the special characters appropriately.

```
In [9]: """The tree's height is 6'2"."""
Out[9]: 'The tree\'s height is 6\'2".'
```

The most common operators are listed in this box:

+	Addition
-	Subtraction
*	Multiplication
/	Division
**	Exponentiation
//	Integer division
%	Modulo (remainder of integer division)
==	Equal to
!=	Differs from
>	Greater than
>=	Greater than or equal to
<=	Less than or equal to
&, and	Logical and
|, or	Logical or
!, not	Logical not

Different operators have different precedence. For example,

```
In [10]: 2 * 3 ** 3
Out[10]: 54
```

This is interpreted as $2 \cdot 3^3$, rather than $(2 \cdot 3)^3$. However, relying on operator precedence makes the code less readable. You should use parentheses to make the calculation clearer:

```
In [11]: 2 * (3 ** 3)
Out[11]: 54
In [12]: (2 * 3) ** 3
Out[12]: 216
```

If this is your first time programming, you may not be familiar with the modulo operator, which returns the remainder of an integer division:

```
In [13]: 15 % 7
Out[13]: 1
```

that is, $15 = (7 \cdot 2) + 1$ (and as such 1 is the remainder of the integer division $15//7$).

3.3.4 Variable Assignment

When programming, typically you manipulate variables. One can think of a variable as a box that contains a value. To create a new variable, simply assign a value to it:

```
In [1]: x = 5 # assign value 5 to variable x
In [2]: x # display x
Out[2]: 5
```

As you can see, Python uses the equals sign to mean "take whatever value is on the right of the equals sign, and assign it to the variable on the left of the equals sign." Note that this means that we will need another symbol to mean "equality," as the equals sign is used for variable assignment. In Python, you can test for equality with the double equals sign == (e.g., 3 == 2 + 1 yields True).

Whenever you create a variable, its name is stored in the memory. To see which variables you have created in the current session, type who:

```
In [3]: who
Out[3]: x
```

Once you have defined a variable, you can use it to perform operations. Each time, Python will look up the value of the variable, and use it to produce the result:

```
In [4]: x + 3
Out[4]: 8
In [5]: y = 8
In [6]: x + y
Out[6]: 13
```

Each variable contains data of a certain type. This data can be of a basic type such as integers, floats, strings, or Boolean, which we have already encountered, or the more complex data structures introduced below. For example, assign a string:

```
# assign a string
In [7]: x = "The cell grew"
# concatenate two strings
In [8]: x + " and is now larger"
Out[8]: 'The cell grew and is now larger'
```

We cannot, however, perform operations on variables that have different types:

```
In [9]: x + y # x is string, y is integer
-------------------------------------------
TypeError    Traceback (most recent call last)
<ipython-input-8-b50c5120e24b> in <module>()
----> 1 x + y

TypeError: Can't convert 'int' object to str implicitly
```

Python raised an error because we cannot concatenate a string (type str) and an integer (type int). However, if it is sensible, we can convert one data type into another using the functions int()—to integer, str()—to string, float()—to floating-point number, etc.

```
In [10]: x # string
Out[10]: "The cell grew"
In [11]: y # integer
Out[11]: 8
In [12]: x + " " + str(y) + " nm"
Out[12]: 'The cell grew 8 nm'
In [13]: z = "88" # string
In [14]: x + z
Out[14]: 'The cell grew88'
In [15]: y + int(z)
Out[15]: 96
```

One of the main features of Python is *dynamic typing*. This means that the type of a variable is determined when the program runs, so in principle you can assign different data types to the same variable within your program. Static-typed languages (such as C or FORTRAN) require you to specify the type of each variable before using it—trying to assign a string to an integer will result in an error (or undesired results). Python, in contrast, automatically determines the type of a variable for you. You can use the function type to determine the type of a variable:

```
In [17]: x = 2
In [18]: type(x)
Out[18]: int
In [19]: x = "two"
In [20]: type(x)
Out[20]: str
```

3.3.5 Built-In Functions

Python provides many built-in functions that you can use to manipulate and query your variables. Above, we have already used the functions type, str, and int. Let's introduce some more. For example, suppose you want to determine the length of the string stored in variable s:

```
In [1]: s = "a long string"
In [2]: len(s)
Out[2]: 13
```

We used the function len(), which returns the length of the variable specified within the parentheses. Here are a few other simple, built-in functions:

```
In [3]: abs(-3.14) # absolute value
Out[3]: 3.14
In [4]: pow(3, 6) # 3^6
Out[4]: 729
In [5]: print(s) # print value of variable s
        a long string
In [6]: round(3.1415926535, 3) # round to 3 digits
Out[6]: 3.142
In [7]: help(round) # call the help function
```

As you can see, each function name is followed by parentheses, which surround the *arguments* of the function. A function may accept additional (optional or mandatory) arguments. For instance, for the function round we specified the number of digits to return (the default is 0 digits). Calling help(FUNCTION NAME) displays brief documentation for the function.

3.3.6 Strings

You have just seen how Python handles basic built-in types such as integers, floats, Booleans, and strings. In this section, we cover strings in more detail. Python excels at string manipulation, making it very useful for biologists who manipulate nucleotide or amino-acid sequences, process the output of laboratory equipment, or parse database information. Let's start by creating a string and applying a general function:

```
In [1]: astring = "ATGCATG"
# return the length of the string
In [2]: len(astring)
Out[2]: 7
```

Python is object oriented, meaning that all variables are objects containing both the data and useful methods to manipulate the data. Think of a method as a function that is specific to that object (e.g., a string). You can access all the methods associated with an object by typing the name of the object followed by a "dot" (.), and pressing Tab (pressing Tab also provides autocompletion).

```
# press Tab after dot to list methods
In [3]: astring.
```

If you want to know more about a type-specific method, use the help function and provide the method's name along with the corresponding type

```
In [4]: help(astring.find)
Help on built-in function find:
[...]
```

Here are some examples of the many string-specific methods in Python:

```
# replace characters
In [5]: astring.replace("T", "U")
Out[5]: 'AUGCAUG'
# position of first occurrence
In [6]: astring.find("C")
Out[6]: 3
# count occurrences
In [7]: astring.count("G")
Out[7]: 2
In [8]: newstring = " Mus musculus "
# split the string (using spaces by default)
In [9]: newstring.split()
Out[9]: [' Mus', 'musculus ']
# specify how to split
In [10]: newstring.split("u")
Out[10]: [' M', 's m', 'sc', 'l', 's ']
# remove leading and trailing white space
In [11]: newstring.strip()
Out[11]: 'Mus musculus'
```

It is possible to use string methods without first assigning the string to a variable. Simply define the string using quotation marks, followed by a dot and the name of the method:

```
# make uppercase
In [12]: "atgc".upper()
Out[12]: 'ATGC'
# make lowercase
In [13]: "TGCA".lower()
Out[13]: 'tgca'
```

To concatenate strings, you can use the plus sign:

```
In [14]: genus = "Rattus"
In [15]: species = "norvegicus"
# separate with a space
In [16]: genus + " " + species
Out[16]: 'Rattus norvegicus'
```

Notice however that concatenating strings using plus signs is rather slow. You can use the method `join` instead:

```
# join requires a list of strings as input; see below
In [17]: human = ["Homo", "sapiens"]
In [18]: " ".join(human)
Out[18]: 'Homo sapiens'
# specify any symbol as delimiter
In [19]: "->".join(["one", "leads", "2", "the", "other"])
Out[19]: 'one->leads->2->the->other'
```

Be careful not to confuse built-in functions with object-specific methods:

```
In [20]: s = "ATGC"
# call the built-in function "print" on a string
In [21]: print(s)
ATGC
# calling the method "print" returns an error message
# that tells us that strings have no method called print
In [22]: s.print()
---------------------------------------------------------

AttributeError Traceback (most recent call last)
<ipython-input-75-1ebfba4a69ba> in <module>()
----> 1 s.print()

AttributeError: 'str' object has no attribute 'print'
```

Intermezzo 3.1

(a) Initialize the string s = "WHEN on board H.M.S. Beagle, as natura-list".

(b) Apply a string method to count the number of occurrences of the character b.

(c) Modify the command such that it counts both lowercase and upper-case b's.

(d) Replace WHEN with When.

3.4 Data Structures

When programming, you need to organize your data so that you can easily access and manipulate it. Python provides useful built-in data structures—special types of objects meant to contain data organized in a certain way. We are going to see *lists*, containing ordered sequences of elements; *dictionaries*, where each element is indexed by a *key*; *tuples*, a list where we cannot change or reorder the elements; and *sets*, a collection of distinct objects. Learning how to use the right data structure for each problem is one of the most important aspects of programming, as a good choice will simplify your work immensely.

3.4.1 Lists

A list is an ordered collection of values, where each value can appear multiple times. It is defined by surrounding the values, separated by commas, with a pair of square brackets:

```
In [1]: new_list = [] # create an empty list
In [2]: my_list = [3, 2.44, "green", True] # specify
    ↳ values
In [3]: a = list("0123456789") # using the list function
In [4]: a
Out[4]: ['0', '1', '2', '3', '4', '5', '6', '7', '8', '9']
```

To access the elements of a list, you can use the index of the element. Note that Python starts indexing at 0.[3] Let's retrieve some elements from our list using an index:

```
In [5]: my_list[1]
Out[5]: 2.44
In [6]: my_list[0]
Out[6]: 3
```

When you try to access a nonexistent element, you will get an error:

---

3. computingskillsforbiologists.com/zeroindexing

```
In [7]: my_list[4]
...
IndexError: list index out of range
```

We can also use index notation to update a value:

```
In [8]: my_list[0] = "blue"
In [9]: my_list
Out[9]: ['blue', 2.44, 'green', True]
```

You can access sequential elements by using the colon (:) operator. This notation is called slicing—think of taking a slice out of the whole (i.e., extract a sublist). Here are some examples:

```
In [10]: my_list
Out[10]: ['blue', 2.44, 'green', True]
In [11]: my_list[0:1] # elements 0 to 1 (noninclusive)
Out[11]: ['blue']
In [12]: my_list[1:3] # elements 1 to 3 (noninclusive)
Out[12]: [2.44, 'green']
In [13]: my_list[:] # take the whole list
Out[13]: ['blue', 2.44, 'green', True]
In [14]: my_list[:3] # from start to element 3
    ↳ (noninclusive)
Out[14]: ['blue', 2.44, 'green']
In [15]: my_list[3:] # from element 3 to the end
Out[15]: [True]
```

The colon creates a sequence of indices starting at the first number, and ending before the second number (so that 0:1 is 0, 0:3 is 0,1,2, and so on). If you don't specify the first number (or the second), the colon operator returns all the indices starting (ending) at that point. Interestingly, you can use negative numbers to index from the end:

```
In [16]: my_list[-2]
Out[16]: 'green'
```

These are the most useful methods for lists:

append
  Append an element to the end of the list.

```
In [17]: my_list.append(25)
In [18]: my_list
Out[18]: ['blue', 2.44, 'green', True, 25]
```

copy
  Create a copy of the list (more on this in section 4.9.3).

```
In [19]: new_list = my_list.copy()
In [20]: new_list
Out[20]: ['blue', 2.44, 'green', True, 25]
```

clear
  Remove all the elements from the list.

```
In [21]: my_list.clear()
In [22]: my_list
Out[22]: []
```

count
  Count occurrences of a certain element in the list.

```
In [23]: seq = list("TKAAVVNFT")
In [24]: seq.count("V")
Out[24]: 2
```

index
  Return the index corresponding to first occurrence of an element.

```
In [25]: seq.index("V")
Out[25]: 4
```

pop

Remove the last element of the list and return it.

```
In [26]: seq2 = seq.pop()
In [27]: seq
Out[27]: ['T', 'K', 'A', 'A', 'V', 'V', 'N', 'F']
In [28]: seq2
Out[28]: 'T'
```

sort

Sort the elements in place (useful for a list of numbers or characters, but can give unexpected results for other types of data).

```
In [29]: a = [1, 5, 2, 42, 14, 132]
In [30]: a.sort()
In [31]: a
Out[31]: [1, 2, 5, 14, 42, 132]
```

reverse

Reverse the order of the elements in place.

```
In [32]: a.reverse()
In [33]: a
Out[33]: [132, 42, 14, 5, 2, 1]
```

To delete an element (or a series of elements) from a list, you can use the function del:

```
In [34]: del(a[2:3])
In [35]: a
Out[35]: [132, 42, 5, 2, 1]
```

3.4.2 Dictionaries

A dictionary is like an unordered list in which the elements are indexed by *keys*. The principle is the same as in an actual dictionary, where definitions are indexed by words. Python dictionaries are useful when the variables do not have a natural order. They are defined by separating key:value pairs using commas, and surrounding them by curly brackets:

```
# create an empty dictionary
In [1]: my_dict = {}
# dictionaries can contain many types of data
In [2]: my_dict = {"a": "test", "b": 3.14, "c": [1, 2, 3,
    ↳ 4]}
In [3]: my_dict
Out[3]: {'a': 'test', 'b': 3.14, 'c': [1, 2, 3, 4]}
In [4]: GenomeSize = {"Homo sapiens": 3200.0, "Escherichia
    ↳ coli": 4.6, "Arabidopsis thaliana": 157.0}
# a dictionary has no natural order
# i.e., the order of key:value input does not matter
In [5]: GenomeSize
Out[5]:
{'Arabidopsis thaliana': 157.0,
 'Escherichia coli': 4.6,
 'Homo sapiens': 3200.0}
# call a specific key (there is no numbering)
In [6]: GenomeSize["Arabidopsis thaliana"]
Out[6]: 157.0
# add a new value using a key not already present
In [7]: GenomeSize["Saccharomyces cerevisiae"] = 12.1
In [8]: GenomeSize
Out[8]:
{'Arabidopsis thaliana': 157.0,
 'Escherichia coli': 4.6,
 'Homo sapiens': 3200.0,
 'Saccharomyces cerevisiae': 12.1}
# nothing happens if the key:value pair already exists
In [9]: GenomeSize["Escherichia coli"] = 4.6
In [10]: GenomeSize
Out[10]:
{'Arabidopsis thaliana': 157.0,
 'Escherichia coli': 4.6,
 'Homo sapiens': 3200.0,
 'Saccharomyces cerevisiae': 12.1}
# the value is overwritten if the key already exists!
In [11]: GenomeSize["Homo sapiens"] = 3201.1
In [12]: GenomeSize
```

```
Out[12]:
{'Arabidopsis thaliana': 157.0,
 'Escherichia coli': 4.6,
 'Homo sapiens': 3201.1,
 'Saccharomyces cerevisiae': 12.1}
```

You just saw that nothing happens if you add a key:value pair that is already in the dictionary. This behavior is useful when you are reading data that might contain duplicate key:value pairs. However, if the key of a new key:value pair matches a key that is already present in the dictionary, the old value will be overwritten (as in In [11]). As for lists, you can delete an element of the dictionary using the del function.

Dictionaries ship with several built-in methods:

copy

Create a copy of the dictionary.

```
In [13]: GS = GenomeSize.copy()
In [14]: GS
Out[14]:
{'Arabidopsis thaliana': 157.0,
 'Escherichia coli': 4.6,
 'Homo sapiens': 3201.1,
 'Saccharomyces cerevisiae': 12.1}
```

clear

Remove all elements.

```
In [15]: GenomeSize.clear()
In [16]: GenomeSize
Out[16]: {}
```

get

Get the value from a key. If the key is not present, return a default value.

```
In [17]: GS.get("Mus musculus", -10)
Out[17]: -10
```

This function is very useful to initialize a dictionary, or to return a special value when the key is not present.

keys

Create a list containing the keys of the dictionary.

```
In [19]: GS.keys()
Out[19]: dict_keys(['Homo sapiens', 'Escherichia coli',
    ↳ 'Arabidopsis thaliana', 'Saccharomyces cerevisiae'
    ↳ ])
```

values

Create a list containing the values of the dictionary.

```
In [20]: GS.values()
Out[20]: dict_values([3201.1, 4.6, 157.0, 12.1])
```

pop(KEY)

Remove the specified key from the dictionary and return the corresponding value.

```
In [21]: GS.pop("Homo sapiens")
Out[21]: 3201.1
In [22]: GS
Out[22]:
{'Arabidopsis thaliana': 157.0,
 'Escherichia coli': 4.6,
 'Saccharomyces cerevisiae': 12.1}
```

update

This is the simplest way to join two dictionaries.

```
In [23]: D1 = {"a": 1, "b": 2, "c": 3}
In [24]: D2 = {"a": 2, "d": 4, "e": 5}
In [25]: D1.update(D2)
In [26]: D1
Out[26]: {'d': 4, 'e': 5, 'b': 2, 'a': 2, 'c': 3}
```

Note that we have updated dictionary D1 with dictionary D2 (the order matters). Values that were not present in D1 were added. If the key already existed, the value was overwritten (see key a).

3.4.3 Tuples

A tuple contains a sequence of values of any type. They are created by surrounding comma-separated values with round brackets. In contrast to lists, tuples are *immutable*. This means that the values in the object cannot be changed once the tuple is defined.

```
In [1]: my_tuple = (1, "two", 3)
# access elements by indexing or slicing
In [2]: my_tuple[0]
Out[2]: 1
# try to assign new value
In [3]: my_tuple[0] = 33
---------------------------------------------------------------
TypeError                    Traceback (most recent call last)
<ipython-input-32-8edf0f29f533> in <module>()
----> 1 my_tuple[0] = 33

TypeError: 'tuple' object does not support item assignment
```

Given that they cannot be modified, tuples come with only two built-in methods:

```
# count elements in tuple
In [4]: tt = (1, 1, 1, 1, 2, 2, 4)
In [5]: tt.count(1)
Out[5]: 4
# return (first) index of element
In [6]: tt.index(2)
Out[6]: 4
```

By defining a sequence of values as a tuple, you are basically giving the data write protection. This is useful for data that should not change, such as the name and coordinates of a field sample or the name and age of an organism. Tuples are also faster than lists. So if you use a sequence only to iterate through it, a tuple is a good choice.

Tuples that contain immutable objects (i.e., strings, numbers) can also be used as keys in a dictionary:

```
D3 = {("trial", 62): 4829}
```

3.4.4 Sets

Sets are lists with no duplicate entries. They come with special operators for union, intersection, and difference. There are two ways to initialize sets: either you can use curly brackets around comma-separated values, or you can call the function set on a list, thereby removing duplicate values. For example,

```
# create a list
In [1]: a = [5, 6, 7, 7, 7, 8, 9, 9]
# use the set function on the list
In [2]: b = set(a)
# duplicate values have been removed
In [3]: b
Out[3]: {5, 6, 7, 8, 9}
In [4]: c = {3, 4, 5, 6}
# intersection
In [5]: b & c
Out[5]: {5, 6}
# union
In [6]: b | c
Out[6]: {3, 4, 5, 6, 7, 8, 9}
# difference: in b but not in c or in c but not in b
In [7]: b ^ c
Out[7]: {3, 4, 7, 8, 9}
```

Above, we have used the logical operators &, |, and ˆ for the union, intersection, and difference of two sets, respectively. These operations are also available as built-in methods. Furthermore, you can test whether a set is a subset (or superset of another):

```
In [8]: s1 = {1, 2, 3, 4}
In [9]: s2 = {4, 5, 6}
In [10]: s1.intersection(s2)
```

```
Out[10]: {4}
In [11]: s1.union(s2)
Out[11]: {1, 2, 3, 4, 5, 6}
In [12]: s1.symmetric_difference(s2)
Out[12]: {1, 2, 3, 5, 6}
In [13]: s1.difference(s2)
Out[13]: {1, 2, 3}
In [14]: s1.issubset(s2)
Out[14]: False
In [15]: s1.issuperset(s2)
Out[15]: False
In [16]: s1.issubset(s1.union(s2))
Out[16]: True
```

It may be confusing to find that calling a = {} creates an empty dictionary, and not an empty set! (On the other hand, there are only three types of brackets you can easily type, but we need to define four main data types.) To initialize an empty set, use a = set([]).

Summary of Data Structures

To recap,

```
# round brackets --> tuple
In [1]: type((1, 2))
Out[1]: tuple
# square brackets --> list
In [2]: type([1, 2])
Out[2]: list
# curly brackets, sequence of values --> set
In [3]: type({1, 2})
Out[3]: set
# curly brackets, key:value pairs --> dictionary
In [4]: type({1: "a", 2: "b"})
Out[4]: dict
```

You can freely combine different data types to create a data structure that fits your data. Just for practice, let's create and access a rather convoluted data structure:

```
In [5]: one = (1, 2, "tuple") # a tuple
In [6]: two = [3, 4, "list"] # a list
In [7]: three = {5: ["value1"], 6: ["value2"]} # a
    ↳ dictionary

# create a list containing the tuple, list, and dictionary
In [8]: container = [one, two, three]
In [9]: container
Out[9]: [(1, 2, 'tuple'), [3, 4, 'list'], {5: ['value1',
    ↳ 'value3'], 6: ['value2']}]

# add a value to the list within the dictionary within the
# list: select the second element of "container" and
# access the list stored under dictionary key "5"
In [10]: container[2][5].append("value3")
In [11]: container
Out[11]: [(1, 2, 'tuple'), [3, 4, 'list'], {5: ['value1',
    ↳ 'value3'], 6: ['value2']}]
```

Intermezzo 3.2

(a) Define a list a = $[1, 1, 2, 3, 5, 8]$.
(b) Extract $[5, 8]$ in two different ways.
(c) Add the element 13 at the end of the list.
(d) Reverse the list.
(e) Define a dictionary m = {"a": ".-", "b": "-...-", "c": '-.-.'}.
(f) Add the element "d": "-..".
(g) Update the value "b": "-...".

3.5 Common, General Functions

Now that we have a better understanding of data structures, we want to mention a few useful, generic functions that are commonly found in programs:

max and min

Return the largest or smallest item in a string, list, or tuple. You can also use max by listing several variables separated by commas (e.g., for integers, floats). If you call the function on a string, it will return the character that has the smallest (largest) associated numerical value. The numerical value is based on the character's lexicographic order (type ord('z') to see which integer 'z' maps into).

```
In [1]: max(a)
Out[1]: 3
In [2]: max(1.2, 3.71, 1.15)
Out[2]: 3.71
In [3]: max("scientific computing")
Out[3]: 'u'
In [4]: min("scientific computing")
Out[4]: ' ' # space has lowest numerical value
```

sum

Return the sum of the elements of a list or a set. Normally, the elements are numbers.

```
In [5]: sum(a)
Out[5]: 6
In [6]: sum(set([1, 1, 2, 3, 5, 8]))
Out[6]: 19
```

in

Test for membership. The expression x in s returns True if x is a member of s, and False otherwise. It works with strings (Is x a substring of s?), lists, tuples, and sets (Is x an element of s?).

```
In [7]: "s" in "string"
Out[7]: True
In [8]: 36 not in [1, 2, 36]
Out[8]: False
In [9]: (1, 2) in [(1, 3), (1, 2), 1000, 'aaa']
Out[9]: True
# for dictionaries, you can test whether a key exists
In [10]: "z" in {"a": 1, "b": 2, "c": 3}
Out[10]: False
In [11]: "c" in {"a": 1, "b": 2, "c": 3}
Out[11]: True
```

3.6 The Flow of a Program

In its simplest form, a program is just a series of instructions (statements) that the computer executes one after the other. In Python, each statement occupies one line (i.e., it is terminated by a newline character). Other programming languages use special characters to terminate statements (e.g., ; is used in C).

3.6.1 Conditional Branching

You can modify the linear flow of a program using special commands. The first command we are going to see creates a branching point. If a certain condition is met, one or more statements are executed, otherwise other statements may be executed.

In this introduction, we are going to type the program directly in Jupyter.[4] Open a new notebook, and title it Conditional. Save it, which will create the file Conditional.ipynb. In the first cell, type

```
x = 4

if x % 2 == 0:
    print("Divisible by 2")
```

Run the commands by pressing Shift+Enter: you should see the output Divisible by 2.

Note the colon (:) at the end of the if statement: you will find it in all commands that alter the flow of a program (e.g., if, else, while, for). Also, the white space before the print statement is very important. These spaces, called "indentation," tell Python which lines need to be executed if the condition is True. Indentation is one of the most controversial aspects of Python (many other programming languages use brackets to mark conditional levels and coherent blocks of code). Throughout the book, we use 4 spaces for each level of indentation (the use of the Tab key is discouraged as its length is not well defined, although you can set up your text editor such that it will print 4 spaces whenever you hit Tab—Jupyter does it automatically). In Python, all the code at the same indentation level is considered to be part of the same block of code.

---

4. Alternatively, you can type the programs in a text editor and run them within Jupyter by calling %run FileName.py.

Now change x = 4 to x = 3, and run it again. The print statement is not executed, as 3 % 2 is 1. Let's extend our program a little:

```
x = 4

if x % 2 == 0:
    print("Divisible by 2")
else:
    print("Not divisible by 2")
```

and try running it with x = 4 and x = 3. The else statement is executed if the condition of the if statement is not met. There can only be one else in combination with an if.

In some cases, we might want to check multiple conditions. Instead of nesting several if statements, we can use an elif statement. As soon as one of the elif conditions is met, the statements are executed, and subsequent statements are ignored. If neither the if nor elif conditions are met, the else statement is executed:

```
x = 17

if x % 2 == 0:
    print("Divisible by 2")
elif x % 3 == 0:
    print("Divisible by 3")
elif x % 5 == 0:
    print("Divisible by 5")
elif x % 7 == 0:
    print("Divisible by 7")
else:
    print("Not divisible by 2, 3, 5, 7")
```

Try setting x to 12 (divisible by 2 and 3): Which statement is executed? In summary, the structure of a conditional branching point is

```
if condition_is_true:
    execute_commands
elif other_condition_is_true:
```

```
    other_commands
else:
    commands_to_run_if_none_is_true
```

3.6.2 Looping

Cycles (or loops) are important modifiers of the flow of a program. These are used to repeat a piece of code multiple times (possibly with slight variations). Examples include iterating through all the files in a directory, all the elements in a list, and all the replicates of an experiment.

Python is equipped with two ways of looping: while, which is used to repeat a piece of code so long as the specified condition is met, and for, which is used to iterate through the elements of a sequence. Here are some examples of while loops:

```
# print the integers from 0 to 99
x = 0
while x < 100:
    print(x)
    x = x + 1
```

```
# print the first few Fibonacci numbers
a = 1
b = 1
c = 0
while c < 10000:
    c = a + b
    a = b
    b = c
    print(c)
```

```
# beware of infinite loops! in Jupyter, to stop
# execution click Kernel -> Interrupt;
# when running Python in terminal, press Ctrl+C
```

```
a = True
while a:
    print("Infinite loop")
```

We can use two further commands to modify the behavior of the loop: break, which stops the cycle, and continue, which skips the remaining code within the loop and moves to the next iteration.

```
# find the first integer >= 15000 that is divisible by 19
x = 15000
while x < 19000:
    if x % 19 == 0:
        print(str(x)  + " is divisible by 19")
        break
    x = x + 1
```

```
# list the first 100 even numbers
x = 0
found = 0
while found < 100:
    x = x + 1
    if x % 2 == 1:
        continue
    print(x)
    found = found + 1
```

The for statement is used for looping through the elements of a sequence (e.g., a list, a string, a tuple, the keys of a dictionary) in the order in which the elements appear in the sequence. For example,

```
# print the elements of a list
z = [1, 5, "mystring", True]
for x in z:
    print(x)
```

```
# print the characters of a string
my_string = "a given string"
for character in my_string:
    print(character)
```

Here is a neat trick to print the key:value pairs of a dictionary. We use the dictionary method items, which returns a list of tuples. The tuples, in turn, contain the key:value pairs of the dictionary. We can use a for loop to iterate through the list of tuples and retrieve the key and value for each entry:

```
# print the keys and values of a dictionary
In [1]: z = {0: "a", 1: "b", 2: "c"}
In [2]: z.items()
Out[2]: dict_items([(0, 'a'), (1, 'b'), (2, 'c')])
In [3]: for (key, val) in z.items():
    print(key, "->", val)
```

Note how we used a tuple to access the key:value pairs of the dictionary. The parentheses around the tuple can be omitted, but are shown for clarity.

The range function is useful when you need to iterate through a range of numbers in a for loop. It creates a Python object of type range. To investigate its behavior, you can convert its output into a list:

```
In [1]: list(range(10))
Out[1]: [0, 1, 2, 3, 4, 5, 6, 7, 8, 9]
In [2]: list(range(1, 5))
Out[2]: [1, 2, 3, 4]
In [3]: list(range(0, 10, 3))
Out[3]: [0, 3, 6, 9]
```

The range function accepts up to three integer parameters. The stop parameter is obligatory and denotes the number up to which the sequence is generated (not including the stop value itself—remember that Python starts counting from 0!). The start and step parameters are optional and denote a possible start other than 0, and the step size of the sequence, respectively. For example,

```
# print x^2 for x in 0 to 9
for x in range(10):
    print(x ** 2)
```

Sometimes, you want to access both an element of a list and its index. Python has a special function, enumerate, that does just that by creating a list of tuples. We can use the list function to look at its output:

```
list(enumerate(my_string))
```

Use enumerate in a loop to print each element of a sequence along with its position:

```
for k, x in enumerate(my_string):
    print(k, x)
```

We can apply enumerate to lists too:

```
z = [1, 5, "mystring", True]
for element, value in enumerate(z):
    print("element: " + str(element) + " value: " +
        ↳ str(value))
```

Finally, if you want to apply the same function to all elements of a list, you can use *list comprehension*, producing very compact code:

```
a = [1, 2, 5, 14, 42, 132]
b = [x ** 2 for x in a]
# this means,
# for each element x in list a, calculate x^2
# and append the result to a new list, called b
print(b)
# [1, 4, 25, 196, 1764, 17424]
```

Intermezzo 3.3

Understanding loops is fundamental to writing efficient programs. Take a moment to consolidate what you have learned and determine how many times "hello" will be printed in the following small programs. Try to come up with the answer before running the code in Python:

(a)
```python
for i in range(3, 17):
    print("hello")
```

(b)
```python
for j in range(12):
    if j % 3 == 0:
        print("hello")
```

(c)
```python
for j in range(15):
    if j % 5 == 3:
        print("hello")
    elif j % 4 == 3:
        print("hello")
```

(d)
```python
z = 0
while z != 15:
    print("hello")
    z = z + 3
```

(e)
```python
z = 12
while z < 100:
    if z == 31:
        for k in range(7):
            print("hello")
    elif z == 18:
        print("hello")
    z = z + 1
```

```
(f)   for i in range(10):
          if i > 5:
              break
          print("hello")
```

```
(g)   z = 0
      while z < 25:
          z = z + 1
          if z % 2 == 1:
              continue
          print("hello")
```

3.7 Working with Files

You will need to open files for input, and save the results of your work in output files. First we cover the use of text files, and then move to more structured, comma-separated-values files (.csv).

3.7.1 Text Files

To read or write a file, first you need to open it:

```
In [1]: f = open("mytextfile.txt", "w")
```

where mytextfile.txt is the path to the file we want to write (relative to where the program is running), and the w stands for "writing" (put r for reading, a for appending). Now f is a file object, also called a *file handle*. You can think of the file handle as the connection between Python and the file. It does not contain the file content, but allows you to access different file methods and properties. Type the name of the file object followed by a dot . and the name of the property:[5]

---

5. Type a dot and hit Tab to see all methods and properties that are available to the object.

```
In [2]: f.name
Out[2]: 'mytextfile.txt'
In [3]: f.mode
Out[3]: 'w'
In [4]: f.encoding
Out[4]: 'UTF-8'
```

The property `encoding` tells us the way the content of the file is interpreted. Basically, files are long "binary" strings of 1s and 0s, and an encoding[6] is a way to map these binary numbers into actual characters. UTF-8 is the standard encoding for the latest versions of Python.

Besides these properties, each file object also comes with specific methods to read and write strings, and to close the file.

There are several methods to access the content of a file. Some read the entire file content at once while some read line by line (better for large files), some methods return the file content as a list, others as a string. Choose a method from the following box that best fits your needs:

read	Return the whole file content as a single string (unless an optional numeric argument is given, specifying how many bytes to return).
readline	Return the current line as a string ending with the newline character \n.
readlines	Return a list of strings, with each element being the corresponding line in the file.

When a file `f` is open for writing, you can

```
f.write(s + "\n") # write string s and go to a new line
f.writelines(["A\n", "B\n", "C\n"]) # write multiple
    ↳ strings provided as a list
```

When you are done reading or writing, you should close the file:

```
f.close()
```

---

6. See computingskillsforbiologists.com/encoding.

Forgetting to close a file is quite common, and can have unintended consequences (e.g., other programs not being able to access the file). To avoid the problem altogether, use `with`:

```
with open("myfile.txt", "r") as f:
    # do some operations, like printing every line
    for my_line in f:
        print(my_line)
```

Using a `with` statement automatically opens and closes the file, preventing mistakes. In the code above, you can also see that you can iterate through the lines of a file using a simple `for` loop. For example, this code reads a file and writes it line by line into another file:

```
inputfile = "test1.txt"
outputfile = "test2.txt"

with open(inputfile, "r") as fr:
    with open(outputfile, "w") as fw:
        # iterate over the lines of a file
        # and write them to another file
        for line in fr:
            fw.write(line)
```

Note that reading is done progressively: once you read a line, you will be able to access the next line, and so on. If you want to go back to the beginning of the file (e.g., if you need to read it twice), use

```
f.seek(0)
```

and it will be as if you just opened the file for reading/writing.

If you want to skip over a line, such as the header of a file, use the method `next`:

```
f.next()
```

3.7.2 Character-Delimited Files

The basic Python installation includes a number of "modules" (covered in the next chapter) providing functions for specific contexts. One of these modules is useful for manipulating character-delimited text files containing values arranged in rows and columns, like in a spreadsheet. These "tabular" text files have two special features: (a) each line is a row of values, separated by a delimiter; (b) each row has the same number of elements. These files are typically called "comma-separated-values" or .csv files, though the separator does not need to be a comma.

To be able to manipulate these files, we need to *import* the module csv. Then we can open a file (for reading or writing) and create a CSV reader (or writer) object. For example, from our CSB/python/sandbox we want to read the first few lines of the file Dalziel2016_data.csv:

```
with open("../data/Dalziel2016_data.csv") as f:
    # create iterator
    for i, line in enumerate(f):
        # print each line; delete leading/trailing spaces
        print(line.strip())
        if i > 2:
            break
```

```
biweek,year,loc,cases,pop
1,1906,BALTIMORE,NA,526822.1365
2,1906,BALTIMORE,NA,526995.246
3,1906,BALTIMORE,NA,527170.1981
```

As you can see, this is a comma-separated file with five columns, and the first line specifies the header (names of the columns). Instead of reading it using the methods introduced above for text files, we can use the csv module:

```
import csv
with open("../data/Dalziel2016_data.csv") as f:
    reader = csv.DictReader(f)
```

```
for i, row in enumerate(reader):
    # print as dictionary
    print(dict(row))
    if i > 2:
        break
```

```
{'cases': 'NA', 'year': '1906', 'pop': '526822.1365', '
    ↳ biweek': '1', 'loc': 'BALTIMORE'}
{'cases': 'NA', 'year': '1906', 'pop': '526995.246', '
    ↳ biweek': '2', 'loc': 'BALTIMORE'}
{'cases': 'NA', 'year': '1906', 'pop': '527170.1981', '
    ↳ biweek': '3', 'loc': 'BALTIMORE'}
{'cases': 'NA', 'year': '1906', 'pop': '527347.0136', '
    ↳ biweek': '4', 'loc': 'BALTIMORE'}
```

Using csv.DictReader(f) to read the file, each row is converted into a dictionary, with keys created automatically from the header: very handy!

We can now perform operations using our file, for example selecting all entries for the location Washington and writing them to a new file using the csv.DictWriter:

```
with open("../data/Dalziel2016_data.csv") as fr:
    reader = csv.DictReader(fr)
    header = reader.fieldnames # extract the header
    with open("Dalziel2016_Washington.csv", "w") as fw:
        writer = csv.DictWriter(fw, fieldnames = header,
            ↳ delimiter = ",")
        for row in reader:
            if row["loc"] == "WASHINGTON":
                writer.writerow(row)
```

Besides csv.DictReader and csv.DictWriter to write data organized in dictionaries, you can also use csv.reader and csv.writer to read and write data as lists.[7]

---

7. For extensive documentation on the module csv, see computingskillsforbiologists .com/csv.

Intermezzo 3.4
Write code that prints the `loc` and `pop` for all the rows in the file `Dalziel 2016_data.csv`.

3.8 Exercises

Here are some practical tips on how to approach the Python exercises (or any programming task):

(a) Think through the problem before starting to write code: Which data structure would be more convenient to use (e.g., sets, dictionaries, lists)?

(b) Break the task down into small steps (e.g., read file input, create and fill data structure, output).

(c) For each step, describe in plain English what you are trying to do—leave these notes as comments within your program to document your code.

(d) When working with large files, initially use only a small subset of the data; once you have tested your code thoroughly you can run it on the whole data set.

(e) Consider using specific modules (e.g., use the `csv` module to parse each line into a dictionary or a list).

(f) Skim through appropriate sections above to refresh your memory on data-type-specific methods.

(g) Use the documentation and help forums.

To maximize your learning, we encourage you to start with an empty Jupyter document, and follow the steps above to solve the exercises. If you are stuck and need some help, then for each exercise, in the directory `CSB/python/solutions` we provide a `.ipynb` document containing pseudocode. Once you have solved the problem, compare your code with the full solution that we present in the same directory.

3.8.1 Measles Time Series

In their article, Dalziel et al. (2016) provide a long time series reporting the numbers of cases of measles before mass vaccination, for many US cities. The data consist of cases in a given US city for a given year, and a given biweek of the year (i.e., first two weeks, second two weeks, etc.). The time series is contained in the file `Dalziel2016_data.csv`.

1. Write a program that extracts the names of all the cities in the database (one entry per city).
2. Write a program that creates a dictionary where the keys are the cities and the values are the number of records (rows) for that city in the data.
3. Write a program that calculates the mean population for each city, obtained by averaging the values of pop.
4. Write a program that calculates the mean population for each city and year.

3.8.2 Red Queen in Fruit Flies

Singh et al. (2015) show that, when infected with a parasite, the four genetic lines of *D. melanogaster* respond by increasing the production of recombinant offspring (arguably, trying to produce new recombinants able to escape the parasite). They show that the same outcome is not achieved by artificially wounding the flies. The data needed to replicate the main claim (figure 2 of the original article) is contained in the file Singh2015_data.csv.

Open the file, and compute the mean RecombinantFraction for each *Drosophila* Line, and InfectionStatus (W for wounded and I for infected). Print the results in the following form:

```
Line 45 Average Recombination Rate:
W :  0.187
I :  0.191
```

3.9 References and Reading

Books

Martin Jones, *Python for Biologists*.
    Available as a set of two books (novice and advanced) and online at pythonforbiologists.com. It contains tools and exercises specifically designed for biologists.
Allen B. Downey, *Think Python: How to Think Like a Computer Scientist*.
    Available online at computingskillsforbiologists.com/thinkpython and as a book by O'Reilly. Explains the basics of computer programming and provides many exercises and case studies.

Documentation and Tutorials

The official Python documentation with detailed explanation and examples: docs.python.org.

Hands-on tutorials in small units: codecademy.com/tracks/python.

On-site workshops in many locations and online lessons: software-carpentry.org/lessons.

CHAPTER 4

• • • • • • • • • • •

Writing Good Code

4.1 Writing Code for Science

When programming for science, you need to make sure that your programs do exactly, and exclusively, what they are meant to do. Bugs (i.e., errors) are not simply annoying, unwanted features (as if you were programming a game app for your phone): any bug in your code can make the results and conclusions of your research unwarranted.

In this chapter we introduce tools that help ensure that your code is correct. First, we provide a few guidelines on how to write your code such that it is easy to understand (for you, and for others), well commented, and easy to debug. We show how to approach a complex program by decomposing it into fundamental building blocks called functions: smaller, coherent blocks of code that can then be organized into a master program.

Next, we show how to leverage the vast number of Python packages that are freely available, and meant to facilitate many common programming tasks. We walk you through a detailed example of how to write your own modules and organize your program structure.

A brief section lists common error messages, showing how they can help you find, solve, and even prevent bugs in your programs. You will also learn how to use a debugger—the favorite tool of proficient programmers.

The section on unit testing introduces a powerful technique that can help you catch the vast majority of problems in your code automatically.

We show how to use a profiler to find the parts of your code where your program is spending most of the execution time. "Optimizing" code for speed, when it is otherwise working correctly, is a major source of errors. Therefore, acting only on the real bottlenecks reduces the risk of introducing bugs.

Finally, we review some more advanced concepts, which give you a glimpse of how Python operates "under the hood."

As always, each of the topics covered in this chapter deserves an entire volume. Here, we focus on the basics and provide a reading list to further your skills on these important aspects of programming.

4.2 Modules and Program Structure

4.2.1 Writing Functions

When writing complex programs, you want to build a clear structure for the flow of the code. The best way to meet this goal is to make your code modular: Write simple building blocks, called functions, each one of which accepts some input, and returns some output. Then combine these blocks to build your program.

Dividing a complex problem into its basic constituents is a bit of an art, with different philosophies. One strategy is to write programs in which each function is very short and performs extremely basic operations. In this case, it is easy to see what each function is doing, but one might lose sight of the overall program if there are very many functions. Alternatively, one can write programs with just a handful of functions, each doing some complex operation. In this case, understanding each function takes more effort, but the flow of the overall program is much simpler.

In the previous chapter, we saw and used many built-in functions (e.g., len, str, print). If there is no function available for a specific task, you can write your own. We start by writing a function that calculates the GC content of a DNA sequence. While this example is not the most efficient way to analyze your sequence data, it nicely illustrates the structure of user-defined functions:

```
# here's our first function
def GCcontent(dna):
    # our function is called GCcontent and
    # accepts a single argument called dna;
    # assume that the input is a DNA sequence encoded
    # in a string, and make sure it's all uppercase:
    dna = dna.upper()
    # count the occurrences of each nucleotide
    numG = dna.count("G")
```

```
numC = dna.count("C")
numA = dna.count("A")
numT = dna.count("T")
# finally, calculate (G + C) / (A + T + G + C)
return (numG + numC) / (numG + numC + numT + numA)
```

The "anatomy" of a Python function is as follows:

- Start with the keyword def followed by the name of the function (in this case, GCcontent).
- Between parentheses, list the inputs of the function (called "arguments") separated by commas, and followed by the colon sign (:).
- All the code belonging to the function is indented.
- The function returns an output, preceded by the keyword return.

Type the code in a new notebook, and run the cell by pressing Shift+Enter. In the next cell, type whos to make sure that your function is among the loaded variables, functions, and modules in your current session:

```
In [2]: whos
Variable Type Data/Info
-------------------------------------------
GCcontent function <function GCcontent at 0x1040c5378>
# the memory address in the third column will differ
```

Now you can use the GCcontent function to calculate the GC content of any DNA sequence:

```
In [3]: GCcontent("AATTTCCCGGGAAA")
Out[3]: 0.42857142857142855
In [4]: GCcontent("ATGCATGCATGC")
Out[4]: 0.5
```

Let's write some other user-defined functions:

```
# print a dictionary
def print_dictionary(mydic):
    for k, v in mydic.items():
```

```
        print("key: ", k, " value: ", str(v))

# return a list with results
# declare default arguments: if no input is provided,
# assume start = 1, end = 10
def squared(start = 1, end = 10):
    # create empty list to catch result of each cycle
    results = []
    for i in range(start, end):
        r = i ** 2
        # append current value to result list
        results.append(r)
    return results
```

Type them in two separate cells in the notebook, and run each cell. You can see that the functions are now available by calling whos:

```
In [7]: whos
Variable Type Data/Info
------------------------------------------------
GCcontent function <function GCcontent at 0x1040c5378>
print_dictionary function <function print_dictionary at
    ↳ 0x1040fe7b8>
squared function <function squared at 0x1040b2048>
```

Now we can call these functions in our programs:

```
In [8]: print_dictionary({"a": 3.4, "b": [1, 2, 3, 4], "c"
    ↳ : "astring"})
key: a value: 3.4
key: b value: [1, 2, 3, 4]
key: c value: astring
```

The function print_dictionary shows that the return statement is optional: you can have a function that does not return any value, but simply performs some operations on the input.

The second function, squared, shows how to store the results of the calculations, returning a list containing all the results. Moreover, the function accepts multiple arguments (start and end), and each argument has a default value, which will be used if the user does not provide it in the function call. For example,

```
# specify both start and end
In [9]: squared(start = 3, end = 10)
Out[9]: [9, 16, 25, 36, 49, 64, 81]
# specify only start, end has default value 10
In [10]: squared(5)
Out[10]: [25, 36, 49, 64, 81]
# specify only end, start has default value 1
In [11]: squared(end = 3)
Out[11]: [1, 4]
# start has default value 1, end is 10
In [12]: squared()
Out[12]: [1, 4, 9, 16, 25, 36, 49, 64, 81]
```

You might have noticed above (e.g., In [10]) that naming the arguments in the function call is optional. The order, however, matters! If we provide only one number, it will be interpreted as the value for the first argument (start); given that we don't provide a second number, the end argument is automatically set to the default value. If we want to omit the start value (as in In [11]), we have to specify that we want to set only the end argument. In general, calling all the arguments of a function by name in your code will make it much more readable, and easier to debug.

Intermezzo 4.1

Functions are another essential building block of programs. Here's a series of small functions: determine what each function does.

(a)
```
def foo1(x = 7):
    return x ** 0.5
```

(b)
```
def foo2(x = 3, y = 5):
    if x > y:
        return x
    return y
```

(c)
```python
def foo3(x = 2, y = 0, z = 9):
    if x > y:
        tmp = y
        y = x
        x = tmp
    if y > z:
        tmp = z
        z = y
        y = tmp
    if x > y:
        tmp = y
        y = x
        x = tmp
    return [x, y, z]
```

(d)
```python
def foo4(x = 6):
    result = 1
    for i in range(1, x + 1):
        result = result * i
    return result
```

(e)
```python
def foo5(x = 1729):
    d = 2
    myfactors = []
    while x > 1:
        if x % d == 0:
            myfactors.append(d)
            x = x / d
        else:
            d = d + 1
    return myfactors
```

(f)
```
# foo6 is a recursive function, meaning that the
# function calls itself;
# read about recursion at
# computingskillsforbiologists.com/recursion
def foo6(x = 25):
    if x == 1:
        return 1
    return x * foo6(x - 1)
```

(g)
```
def foo7(x = 100):
    myp = [2]
    for i in range(3, x + 1):
        success = False
        for j in myp:
            if i % j == 0:
                success = True
                break
        if success == False:
            myp.append(i)
    return myp
```

4.2.2 Importing Packages and Modules

A *module* is a single file that contains a collection of functions. A *package* is an organized collection of modules. Python ships with many packages, which you can import into your code to gain access to many functions. These packages are freely available and can save you a lot of programming—you do not have to reinvent the wheel! In most cases, packages have also been thoroughly tested and, therefore, tend to be very reliable. For example, the module os allows you to interface with the operating system (e.g., create directories, check whether a file exists) and the re module is devoted to regular expressions (which we will cover in chapter 5); many more are available.[1]

---

1. For a short list, see computingskillsforbiologists.com/pythonmoduleindex.

How can you access the functions in a module? There are four different ways to load the module `mymodule` in order to access the function `my_function` within the module:

`import mymodule`

Import the complete module `mymodule`. A specific function in the module can be accessed by typing `mymodule.my_function()`. This is the preferred way to import modules, as specifying the module name every time you call a function means that you will not lose track of where the function came from (and works even if you are importing two modules that contain functions with the same name!).

`from mymodule import my_function`

In this way, you import only the function `my_function` contained in module `mymodule`. The function can then be accessed as if it were part of the current program: `my_function()`.

`import mymodule as mm`

Import the complete module `mymodule` and call it `mm`. This is convenient only if the name of the module is very long. The functions in the module can be accessed as `mm.my_function()`.

`from mymodule import *`

Import all the functions in module `mymodule` as if they were part of the current program. This should be avoided, as it can generate *namespace pollution*: that is, by importing all functions and variables from a module, you overwrite current functions and variables that have the same name.

You can (and should) create your own modules! Simply save the functions in a file (e.g., `genesmod.py`) and place the file in a directory where your other programs can access it (possibly in the same directory). Then, you can type `import [NAME_OF_FILE]` (e.g., `import genesmod`) to access all your functions (e.g., `genesmod.name_of_function()`). In fact, as we will see below, this is the best way to organize your code.

4.2.3 Program Structure

So far, we've been writing short scripts that perform a handful of operations. When writing more substantial programs, it is important to subdivide the work into manageable pieces that are easier to test, debug, and maintain.

To practice coding using a more complex program structure, we are going to write a simulation dealing with population genetics. We want to simulate a population of N monoecious (i.e., hermaphrodites), diploid (i.e., carrying two homologous copies of each chromosome) organisms. We are focusing on a particular gene, which has two alternative forms (alleles), *A* and *a*. Initially, the individuals are assigned a genotype, receiving allele *A* with probability *p*, and allele *a* with probability 1 − *p*. At each generation, the organisms reproduce, and then die (nonoverlapping generations). For simplicity, we assume that the population size is constant, that there are no mutations, that there is no selection (i.e., the different genotypes have the same fitness), and that mating is completely random (for each offspring, we choose two random parents).

Then we use the simulation to explore the concept of genetic drift: for small populations, even alleles that do not bring a fitness advantage can go to fixation (i.e., be present in 100% of the individuals).

Writing a complex simulation might seem daunting at first, but dividing the program into its basic constituents (i.e., separate functions) makes it much easier. A "master" program will then call the functions and produce the desired result. Before typing any code, we sketch a possible division into functions, writing in plain English:

- A function that initializes the population: It should take as input the size of the population (*N*), and the probability of having an *A* allele (*p*). This function returns an entire population.
- A function that computes the genotypic frequencies, which we will need to determine whether an allele has gone to fixation: The function should take a population as input and output the count for each genotype.
- A reproduction function that takes the current population and produces the next generation.

We also need to choose a data structure for our program. Here we represent a population as a list of tuples, where each tuple is an individual with its two chromosomes (e.g., ("A", "A") would be a homozygous individual).

We start by importing the module SciPy. This module (covered in chapter 6) contains many useful scientific functions. Here we use it only to draw random numbers. We write the first function, which generates a population:

```
import scipy  # for random numbers

def build_population(N, p):
    """The population consists of N individuals.
        Each individual has two chromosomes, containing
```

```
        allele "A" or "a", with probability p or 1-p,
        respectively.

        The population is a list of tuples.
    """
    population = []
    for i in range(N):
        allele1 = "A"
        if scipy.random.rand() > p:
            allele1 = "a"
        allele2 = "A"
        if scipy.random.rand() > p:
            allele2 = "a"
        population.append((allele1, allele2))
    return population
```

Save the notebook. We can test that the function works by pressing Shift+Enter, and then calling the function in the next cell:

```
In [3]: build_population(N = 10, p = 0.7)
Out[3]: [('a', 'A'),
         ('A', 'A'),
         ('a', 'a'),
         ...
```

Note that your output might look different, as we are building the population at random! Now we write the second function, which is used to produce a genotype count for the population:

```
def compute_frequencies(population):
    """ Count the genotypes.
        Returns a dictionary of genotypic frequencies.
    """
    AA = population.count(("A", "A"))
    Aa = population.count(("A", "a"))
    aA = population.count(("a", "A"))
    aa = population.count(("a", "a"))
```

```
return({"AA": AA,
        "aa": aa,
        "Aa": Aa,
        "aA": aA})
```

Let's see the function at work:

```
In [5]: my_pop = build_population(6, 0.5)
In [6]: my_pop # this might be different---random values!
Out[6]: [('a', 'A'), ('a', 'A'), ('a', 'A'), ('A', 'A'),
    ↳ ('A', 'a'), ('A', 'a')]
In [7]: compute_frequencies(my_pop)
Out[7]: {'Aa': 2, 'aA': 3, 'aa': 0, 'AA': 1}
```

Finally, let's write the most complex function, which produces a new generation of the population:

```
def reproduce_population(population):
    """ Create new generation through reproduction
        For each of N new offspring,
        - choose the parents at random;
        - the offspring receives a chromosome from
            each of the parents.
    """
    new_generation = []
    N = len(population)
    for i in range(N):
        # random integer between 0 and N-1
        dad = scipy.random.randint(N)
        mom = scipy.random.randint(N)
        # which chromosome comes from mom
        chr_mom = scipy.random.randint(2)
        offspring = (population[mom][chr_mom], population
            ↳ [dad][1 - chr_mom])
        new_generation.append(offspring)
    return(new_generation)
```

Again, we perform a quick test to see that everything works as expected:

```
In [9]: reproduce_population(my_pop)
Out[9]: [('A', 'A'), ('A', 'a'), ('a', 'a'), ('A', 'A'),
    ↳ ('a', 'a'), ('a', 'A')]
```

Now that we have all the pieces in place, we need to build our main program. We have two choices: either write a general function at the end of this code, or write a second notebook and import the functions we wrote as a module. Here we adopt the second strategy, which is very convenient when you have many functions. Unfortunately, Jupyter makes it a little difficult to import notebooks directly.[2] Thus, we are going to (1) export the notebook as a flat Python file (.py), and (2) import the module into a second notebook. First, remove all the code you ran for tests by clicking "Edit" → "Delete Cells" in the Jupyter top panel (i.e., keep only the cells containing the definitions of the functions). Then choose "Download As" and Python (.py) from the "File" menu. Open the file you just downloaded and save it into your sandbox as drift.py. (For your convenience, you will find a complete copy of the module in CSB/good_code/solutions/drift.py.) Now launch a new notebook, called simulate_drift, and in the first cell, type

```
In [1]: import drift
```

You can test whether the import was successful by typing help(drift). The automatically generated help file lists all available functions. Note that if you start a function by including a short description flanked by triple quotes (i.e., what Python calls a *docstring*), this will be used to generate the help page. You should always document your functions in this way.

Now that you have imported the module drift, you can access all the functions in it by typing drift. and pressing Tab.

In the next cell, write the main code:

```
def simulate_drift(N, p):
    # initialize the population
    my_pop = drift.build_population(N, p)
    fixation = False
    num_generations = 0
    while fixation == False:
        # compute genotype counts
```

---

2. For a work-around, see computingskillsforbiologists.com/importjupyter.

```
        genotype_counts = drift.compute_frequencies
            ↳ (my_pop)
        # if one allele went to fixation, end
        if genotype_counts["AA"] == N or genotype_counts["
            ↳ aa"] == N:
            print("An allele reached fixation at
                ↳ generation", num_generations)
            print("The genotype counts are")
            print(genotype_counts)
            fixation == True
            break
        # if not, reproduce
        my_pop = drift.reproduce_population(my_pop)
        num_generations = num_generations + 1
```

Notice that the program is very easy to read. Another scientist would have to understand only this file to get a sense of what the code is doing. Let's try out our new simulation:

```
In [3]: simulate_drift(100, 0.5)
An allele reached fixation at generation 66
The genotype counts are
{'aa': 100, 'aA': 0, 'AA': 0, 'Aa': 0}
In [4]: simulate_drift(100, 0.9)
An allele reached fixation at generation 20
The genotype counts are
{'aa': 0, 'aA': 0, 'AA': 100, 'Aa': 0}
```

Note that your numbers might vary, as the simulation involves randomness. If you want to store your randomly created population, so you can later work with the exact same version, you can use the module pickle. You can *pickle* Python objects, such as tuples, lists, sets, and dictionaries, which turns them into a byte stream that can later be *unpickled*. Here is an example:

```
import pickle
# save Python object to file
pickle.dump(my_pop, open("population.pickle", "wb"))
```

Note that the file extension .pickle is an arbitrary, yet informative choice. Opening the file in a text editor will give you only gibberish. Here is an example of how to load the data back into your Python session:

```
population = pickle.load(open("population.pickle", "rb"))
print(population)
```

Beware of unpickling code from unknown sources as you may load malicious code.

4.3 Writing Style

As the adage goes, code is read more often than it is written. It is good practice to think of your readers while writing your code: Can another scientist understand what you are doing, how you are doing it, and, possibly, why? Even more importantly, will *you* be able to understand your own code six months from today? This is not a purely academic question, given that it takes a long time to publish a paper. When the reviews are back and you need to modify your code, having it well documented and organized will make your life simpler.

The following guidelines are meant to increase the readability of your code:

- Use four spaces per indentation level; not tabs.
- Split long lines (especially the arguments of functions) using parentheses:

```
# yes:
def my_function_name(argument1,
                     argument2,
                     argument3,
                     argument4,
                     argument5):
    one_operation()
    another_operation()
    if condition:
        yet_another_operation()
    return my_value
```

```
for each_item in my_list:
    for each_element in my_item:
        do_something()
# no:
def my_function_name(
    ↳ argument1,argument2,argument3,argument4,
    ↳ argument5):
 one_operation()
 another_operation()
 if condition:
        yet_another_operation()
 return my_value

for each_item in my_list:
  for each_element in my_item:
            do_something()
```

- Separate functions using blank lines.
- Use a new line for each operation:

```
# yes:
import scipy
import re
# no:
import scipy,re
```

- Import packages and modules at the beginning of the file, in this order:
    1. Standard modules (i.e., those shipped with Python)
    2. Third-party modules (i.e., those developed by other groups)
    3. Modules you have developed
- Put spaces around operators (unless using the : to "slice" a list); use parentheses to clarify the precedence of operators:

```
# yes:
a = b ** c
z = [a, b, c, d]
```

```
n = n * (n - 1)
k = (a * b) + (c * d)
my_list = another_list[1:n]
# no:
a=b**c
z=[ a,b,c,d ]
n =n*(n-1)
k=a*b+c*d
my_list = another_list[1 : n]
```

- Python allows you to write short comments (starting with #) and *docstrings* (multiline comments flanked by triple double quotes, """):
  - Use *docstrings* to document how to use the code. What does this function do? What is the input/output?
  - Use short comments for explaining difficult passages in the code, that is, to document how the code works.
- Naming conventions: Variables and functions are in lowercase (a_variable, my_function), constants are in uppercase (SOME_CONSTANT); use underscores to separate words. Use descriptive names suggestive of what the function does or what the variable is (e.g., count_number_pairs, body_mass, cell_volume). Use verbs for functions and nouns for Python objects. Never call a variable o (lowercase O), O (uppercase O), I (uppercase i), or l (lowercase L), as these are easily confused with the digits 0 (zero) and 1 (one).

These are only a few suggestions. For more detailed style guides and a list of pet peeves, see section 4.11 at the end of the chapter.

4.4 Python from the Command Line

Wouldn't it be nice to be able to call our program simulate_drift.py directly from the command line? In that way, it would be very easy to automate the whole pipeline. Notebooks are meant for working in interactive mode but become somewhat cumbersome when you want to call scripts programmatically. However, as we saw above, Jupyter allows you to export your notebooks in plain .py files that can be run by Python without the need to open Jupyter first. Clean the code in your notebook by removing any unnecessary cells, and export the file as simulate_drift.py in your sandbox. Then open the file in a text editor. Now follow these three steps:

1. Add a first line to the code, specifying where python3 can be found.[3] We also import the module sys that lets us capture arguments from the terminal that we want to use within the Python program:

```
#!/usr/bin/python3
import sys
```

2. Add the following code at the end of the file:

```
if __name__ == "__main__":
    # read the arguments on the command line
    # and convert to the right type
    # (they are strings by default)
    N = int(sys.argv[1])
    p = float(sys.argv[2])
    # call the simulation
    simulate_drift(N, p)
```

This special code tells Python that when this is executed by itself (that's what the __name__ == "__main__" stands for), the arguments of the command line should be passed to the function simulate_drift.

3. Make the Python script executable:[4]

```
$ chmod +rx simulate_drift.py
```

Now you can execute your script directly from the command line:

```
$ ./simulate_drift.py 1000 0.1
An allele reached fixation at generation 3332
The genotype counts are
{'aa': 0, 'Aa': 0, 'AA': 1000, 'aA': 0}
```

---

3. If you don't know the path to your Python installation, find out by typing whereis python3 in your Unix terminal or where python3 in Windows Git Bash.

4. See section 1.6.7 to refresh your memory on file permissions.

4.5 Errors and Exceptions

Python distinguishes between two types of errors: syntax errors and exceptions. A SyntaxError is raised when you have not abided by the syntax (i.e., grammar) of Python. Exceptions produce a Traceback message and are raised when the code is grammatically correct, but cannot be executed. Error and traceback messages might look intimidating at first, but they happen to everyone. In fact, these messages are incredibly useful as they point to the position and describe the type of an error. Let's look at some messages that are commonly encountered:

SyntaxError
> Your code contains a syntax (i.e., grammar) error. Common problems are a missing colon after a conditional or loop statement (if, for, etc.), forgetting quotes when defining a string, or using an assignment (=) instead of a (==) comparison. The error is also raised when you try to name a variable using a name that is a Python keyword (for instance, yield, class, etc.).

IndentationError
> The amount of indentation is incorrect, such as after invoking a conditional branching, function, or loop.

TypeError
> Your variable type does not allow the operation you tried to perform on it, such as adding a number and a string.

NameError
> The variable does not exist, either because you did not define it, or due to a typo.

IndexError
> You tried to access a nonexistent item in a list.

KeyError
> You tried to access a nonexistent item in a dictionary.

IOError
> You tried to read a file that does not exist, or tried to write to a file that is open only for reading.

AttributeError
> You tried to access an attribute or method that does not exist, such as .keys() on a list. Use x. and hit Tab to see which attributes and methods are available for variable x.

Besides these, which are very common, Python can return other messages.[5]

4.5.1 Handling Exceptions

When writing your own programs, it can be useful to anticipate errors and prevent them from stopping the execution of your code.

Suppose we want to divide a number y by another, x. However, following the mathematical convention, Python does not know what to do if we try to divide by zero:

```
In [1]: x = 6
In [2]: y = 2.0
In [3]: y / x
Out[3]: 0.3333333333333333
In [4]: x = 0
In [5]: y / x
---------------------------------------------------------
ZeroDivisionError Traceback (most recent call last)
----> 1 y / x
ZeroDivisionError: float division by zero
```

As expected, Python raises an exception and stops. We can catch such an exception and avoid the error message by using special code that runs when the problem is encountered. For example, you can type

```
y = 16.0
x = 0.0

try:
    print(y / x)
except:
    print("Cannot divide by 0")
print("I'm done")
```

Here, we have used except by itself, which catches any problem and executes an appropriate alternative block of code when an exception is found

---

5. You can find a detailed list at computingskillsforbiologists.com/pythonexceptions and a neat cheat sheet at computingskillsforbiologists.com/commonpythonerrors.

(e.g., try to set y = 'abc'). However, if possible, we should define exceptions more specifically:

```
y = 16.0
x = 0.0

try:
    print(y / x)
except ZeroDivisionError:
    print("cannot divide by 0")
print("I'm done")
```

4.6 Debugging

When faced with a bug in their code, many programming novices start adding print statements here and there in the code, hoping to zero in on what is causing the issue. There are much better alternatives that help you locate and identify errors. Python ships with pdb, which is short for Python debugger. You can turn it on in your Jupyter notebook by typing

```
In [1]: %pdb
Automatic pdb calling has been turned ON
```

Calling %pdb again switches the debugger off. For now, keep it switched off.

What is a debugger? This is special software that you can use to follow your code line by line, or to take a look at what is happening at a specific point of the code (called a *breakpoint*). At any moment, you can inspect all variables (e.g., What is the value of my_file when I am executing line 26?) and move to the next line of code. If the debugger is turned on, every time the program encounters an error, instead of quitting, it will enter the debugging mode, and you will have a chance to see what went wrong.

To illustrate the use of the debugger, we will write a piece of code with a bug. Type this code in a new notebook:

```
1   # import a function for normal distribution
2   from numpy.random import normal
3   # import a function for uniform distribution
```

```
4    from numpy.random import uniform
5    # a function to perform the sqrt
6    from math import sqrt
7
8    def get_expected_sqrt_x(distribution = "uniform",
9                            par1 = 0,
10                           par2 = 1,
11                           sample_size = 10):
12       """ Calculate the expectation of sqrt(X)
13       where X is a random variable.
14       X can be either uniform or normal,
15       with parameters specified by the user
16       """
17       total = 0.0
18       for i in range(sample_size):
19           if distribution == "uniform":
20               z = uniform(par1, par2, 1)
21           elif distribution == "normal":
22               z = normal(par1, par2, 1)
23           else:
24               print("Unknown distribution. Quitting...")
25               return None
26           total = total + sqrt(z)
27       return total / sample_size
```

Before we turn to debugging, let's try to understand the code:

Line 2   We import the function normal from the module numpy.
              random. We can now call the function directly (i.e., we do not
              need to type numpy.random.normal, but simply normal). The
              same is done for the functions uniform and sqrt.

Line 8   We set a default value "uniform" for the argument
              distribution. If the user does not specify a value, this value
              will be used. Lines 9–11 define default values for the other
              arguments of the function.

Lines 12–16   We explain what the function does immediately after the func-
              tion definition, using a *docstring*.

Lines 19–25   Notice the use of if … elif …else. If the user does not pro-
              vide a known distribution, the function returns None, a special
              Python type used for lack of a value.

Line 20   The `uniform` function takes three arguments: a minimum (`par1`, 0 by default), a maximum (`par2`, 1 by default), and a value specifying how many numbers should be drawn (1 in this case). The function `normal` is similar, but the first parameter determines the mean and the second parameter the standard deviation.

Now, let's run this code:

```
In [2]: get_expected_sqrt_x(sample_size = 100)
Out[2]: 0.6522005285955806
```

This is quite close to the asymptotic value obtained when the sample size is very large: if X is uniformly distributed between 0 and 1, then the expected value of $\sqrt{X}$ is given by $\mathbb{E}[\sqrt{X}] = \int_0^1 \sqrt{x}\,dx = \frac{2}{3}$. You should have obtained a similar number (slightly different from above, given the randomness involved in the process).

Now let's change the input arguments and use a normal distribution:

```
In [3]: get_expected_sqrt_x(distribution = "normal", par1
     ↳ = 1, par2 = 0.5, sample_size = 10)
Out[3]: 0.9724022318219052
```

Chances are you did not get an error (in fact, this should run without throwing an error with a probability of $(1 - 0.02275)^{10} \approx 0.794$). However, increasing the sample size makes it more likely that you encounter an error:

```
In [4]: get_expected_sqrt_x("normal", 1, 0.5, 1000)
-------------------------------------------
ValueError Traceback (most recent call last)
<ipython-input-12-cfb52ea3a777> in <module>()
----> 1 get_expected_sqrt_x("normal", 1, 0.5, 1000)
[...]thereisabug.py in get_expected_sqrt_x(distribution,
     ↳ par1, par2, sample_size)
    24 print("Unknown distribution. Quitting...")
    25 return None
```

```
---> 26 total = total + sqrt(z)
     27 return total / sample_size

ValueError: math domain error
```

The standard error message shows only that the error happens when trying to execute line 26. Time to turn our debugger on:

```
In [5]: %pdb
Automatic pdb calling has been turned ON
# call the function again with the same values
In [6]: get_expected_sqrt_x("normal", 1, 0.5, 1000)
[...]
ValueError: math domain error
[...]
     25 return None
---> 26 total = total + sqrt(z)
     27 return total / sample_size
ipdb>
```

This looks exactly as before, but now, instead of being sent back to the next cell, we entered the ipdb shell. From there, we can examine values of all variables and run individual lines of code. For instance, we can investigate the value of each variable when the error happened:

```
ipdb> total
50.69164023148862
ipdb> sample_size
1000
ipdb> sqrt(4)
2.0
```

In this case, printing the value of z shows what the problem is:

```
ipdb> z
array([-0.43883187])
```

That is, we are trying to calculate the square root of a negative number, and according to mathematical convention Python cannot execute the function and returns an error. To exit the ipdb shell, type q.

In this case, our code does not make sense. Maybe the best solution is to take the absolute value of z, and document this change in the function:

```python
# import a function for normal distribution
from numpy.random import normal
# import a function for uniform distribution
from numpy.random import uniform
# import a function to perform the sqrt
from math import sqrt

def get_expected_sqrt_abs_x(distribution = "uniform",
                            par1 = 0,
                            par2 = 1,
                            sample_size = 10):
    """ Calculate the expectation of sqrt(|X|)
    where X is a random variable.
    X can be either uniform or normal,
    with parameters specified by the user;
    before taking the square root, we take the
    absolute value, to make sure it's positive.
    """
    total = 0.0
    for i in range(sample_size):
        if distribution == "uniform":
            z = uniform(par1, par2, 1)
        elif distribution == "normal":
            z = normal(par1, par2, 1)
        else:
            print("Unknown distribution. Quitting...")
            return None
        total = total + sqrt(abs(z))
    return total / sample_size
```

In our example, the execution halted. In such a case, debugging is typically quite easy. Turn on your debugger, poke around, and check the value of all the variables; you will find what is causing the problem. Similarly, when a function returns some strange result (e.g., nan, a number instead of a string, variables of the wrong type), it is usually quite easy to find the mistake(s).

You do not need to wait for an error to occur to start debugging. Importing pdb and including the command pdb.set_trace() at any point of your code will make the debugger start at that line (i.e., it will create a breakpoint). In fact, when writing complex functions, the following procedure is good coding practice:

- Write a bit of code.
- Set a breakpoint.
- Run the code and check that everything is fine.
- Within the debugger, try the commands that you want to add to the program and make sure they have the desired results.
- Copy the lines into the function.
- Rinse and repeat.

In general, the code that is the most difficult to debug is that which does not produce an error message but instead returns a reasonable, albeit incorrect, answer. In these cases, bugs can go undiscovered for a long time. You may have heard about some prominent cases (e.g., the string of retractions by Geoffrey Chang's group (Miller, 2006) caused by a minuscule bug that flipped some signs in the data), but the number of scientific programs containing errors is unknown and is likely to be vast.

The best strategy to deal with this problem is to have a clear idea of what the result of a function or a piece of code should be, at least in some simple cases.

Another way to discover and prevent bugs is to use *assertions*. If you know that at a particular place in your code, a certain variable should have a given property, you can "assert it" and Python will raise an exception (i.e., halt) if the property is not satisfied. Assertions are often used to make sure that the input to a function has certain qualities:

```python
import scipy  # for log and exp
import scipy.special  # for binomial coefficient

def compute_likelihood_binomial(p, successes, trials):
    """ Compute the likelihood function for the binomial
        model where p is the probability of success;
```

```
            successes is the number of observed successes
            out of trials
        """
        assert p >= 0, "Negative probability"
        assert p <= 1, "Probability > 1!"
        assert successes <= trials, "More successes than
            ↳ trials!"
        log_likelihood = successes * scipy.log(p)
            + (trials - successes) * scipy.log(1.0 - p)
        return scipy.exp(log_likelihood) * scipy.special.binom
            ↳ (trials, successes)
```

```
In [2]: compute_likelihood_binomial(0.5, 5, 10)
Out[2]: 0.24609375
In [3]: compute_likelihood_binomial(-0.5, 5, 10)
-------------------------------
AssertionError Traceback (most recent call last)
...
----> 9 assert p >= 0, "Negative probability"
     10 assert p <= 1, "Probability > 1!"
     11 assert successes <= trials, "More successes than
            ↳ trials!"

AssertionError: Negative probability
```

Similarly, a few well-placed assertions can make your code more readable and easier to debug.

Intermezzo 4.2

Here is a small program that takes an mRNA sequence, and translates all the codons into the corresponding amino acids. We start at the first nucleotide and consider only one reading frame. The program halts when it encounters a stop codon (UAA, UAG, or UGA). However, there's a bug in the program! Use the debugger to inspect the code, find the problem, and correct it.

For your convenience, the script is available in the directory good_code/ data. Open a new Jupyter notebook in the sandbox, copy the code into a cell, and start debugging.

```
import pickle
# load dictionary with genetic code from pickle file
genetic_code = pickle.load(open("../data/
    ↳ genetic_code.pickle", "rb"))

# test case: desired amino acid sequence
# MEFSL[stop]
test_mRNA = "AUGGAAUUCUCGCUCUGAAGGUAA"

def get_amino_acids(mRNA):
    i = 0
    aa_sequence = []
    while (i + 3) < len(mRNA):
        codon = mRNA[i:(i + 3)]
        aa = genetic_code[codon]
        if aa == "Stop":
            break
        else:
            aa_sequence.append(aa)
        # advance to the next codon
        i = i + 4
    return "".join(aa_sequence)

print(get_amino_acids(test_mRNA))
# problem: the program returns MNLLEV instead of MEFSL!
```

4.7 Unit Testing

The notion of checking that a function is returning the correct answer for test cases has been formalized in the idea of unit testing. Unit testing means writing your code in a way that prevents common mistakes, and encourages writing solid, reliable, and well-documented code. The gist of it is to write *independent* tests for the *smallest units* of code, that is, whenever you write a function, you write a small piece of code to test it. Testing is then done automatically, so that whenever you modify the code, you can make sure all tests are still returning the right value.

Why should we write these tests? The reason is that most bugs are introduced when you make small changes to functions—typically to add a feature you forgot, or to deal with data that are slightly different from what you had in mind. When you make changes to a function, you can run the tests to make sure that the changes have not altered its original behavior. Clearly, writing a set of tests that captures most (better, all) of the important features of your function is what makes this approach useful, and you should spend time thinking about the best ways to test your functions.

What is the best way to use unit testing? First, when you write your code, you divide functions into different files (modules), so that similar functions are in the same file (e.g., one module deals with the statistics, one with the simulations, one with the input). All these modules are imported by the main program, which is then quite short and easy to follow—all the gory details are in the modules. With this organization of the modules, we can then test each module effortlessly and automatically whenever we make the slightest change to the code. If the change causes some tests to fail, you can rest assured you just introduced a bug—go ahead and fix it! Without the tests, you would have continued a happy day without knowing that all your results were wrong.

4.7.1 Writing the Tests

Above we outlined the basic theory of unit testing. In Python, moving from theory to practice is easy, as there are special modules to perform unit testing. Here we work with the simplest one, doctest, which uses *docstrings* to perform unit testing.[6]

It is much easier to write tests in a flat Python file than in Jupyter. For this reason, we are going to work in a text editor, and invoke Python from the command line. Write the function CGcontent (similar to the function we wrote in section 4.2.1) and save it as CGcont.py in CSB/good_code/sandbox:

```python
#!/usr/bin/python3

def CGcontent(DNA):
    """ Return proportion of CG in sequence.
        Assumes that the DNA is uppercase containing
        ONLY A T G C
    """
    CG = DNA.count("C") + DNA.count("G")
    CG = CG / len(DNA)
    return CG
```

---

6. We saw in section 4.3 how to use *docstrings* to document what a function is doing.

Now that we have a function as one small, functional unit, we want to construct some tests. This can be accomplished in three easy steps:

1. Launch python3 (not Jupyter) from the command line, specifying that you want to import the code with the function you want to test:

```
$ python3 -i CGcont.py
```

2. Now write some tests for the function, ideally with parameters spanning the whole range of possibilities:

```
>>> CGcontent("AAAAAA")
0.0
>>> CGcontent("AAATTT")
0.0
>>> CGcontent("AAATTTCCCC")
0.4
>>> CGcontent("AAATTTCCC")
0.33333333
```

Note that the last test yields a number that could be slightly approximated due to rounding errors (i.e., another version of Python might show more or fewer numbers in the fractional part).

3. Finally, copy your test into the *docstring* of the function. You can also include comments to document the test itself:

```
#!/usr/bin/python3


def CGcontent(DNA):
    """ Return proportion of CG in sequence.
        Assumes that the DNA is uppercase containing
        ONLY A T G C
    ===================================
    Unit testing with docstrings
    ===================================
    Run the command in python3, (e.g., python3 -i
        ↳ CGcont.py) and copy the output below:

    >>> CGcontent("AAAAAA")
```

```
0.0
>>> CGcontent("AAATTT")
0.0
>>> CGcontent("AAATTTCCCC")
0.4
>>> CGcontent("AAATTTCCC") # doctest: +ELLIPSIS
0.333...
"""
CG = DNA.count("C") + DNA.count("G")
CG = CG / len(DNA)
return CG
```

The docstring now looks a bit like an interactive session. When we run the unit test, the lines with the leading >>> will be compared to the expected results. If you altered the code of the function itself and introduced a bug, the tests will fail. The comparison is verbatim, that is, if the obtained result does not match the expected result exactly(!), the test fails. We have pointed out that different versions of Python might produce slightly different decimals. This can be solved by using an ellipsis (...) in the test. This signals doctest to use only the first few digits of the output for testing. As shown in the code above, you have to add # doctest: +ELLIPSIS to the end of the line for this to work.

4.7.2 Executing the Tests

You can execute the doctest in your terminal:

```
$ python3 -m doctest -v CGcont.py
Trying:
    CGcontent("AAAAAA")
Expecting:
    0.0
ok
Trying:
    CGcontent("AAATTT")
Expecting:
    0.0
ok
Trying:
```

```
        CGcontent("AAATTTCCCC")
Expecting:
    0.4
ok
Trying:
    round(CGcontent("AAATTTCCC"), 3)
Expecting:
    0.333...
ok
1 items had no tests:
    CGcont
1 items passed all tests:
    4 tests in CGcont.CGcontent
4 tests in 2 items.
4 passed and 0 failed.
Test passed.
```

The option `-m` lets us import our script `CGcont.py` as a stand-alone module and execute `doctest` on it. The option `-v` causes Python to output a detailed description of the test: what was expected and what was returned. Every time you change your module or your functions, you should run the tests again. You can automate the whole procedure by adding these lines to the bottom of the script:

```
if __name__ == "__main__":
    import doctest
    doctest.testmod()
```

Now every time you run the script as a stand-alone piece of code (i.e., by running `python3 CGcont.py -v`), the tests are executed automatically.

4.7.3 Handling More Complex Tests

The module `doctest` makes it very easy to write simple tests where the output of the test can be compared character by character with what was written in the *docstring*. In the following, we deal with cases that require special handling.

The Function Returns an Unordered Object

If your function returns a dictionary, you might fail to pass the test even if the answer is correct: the keys in the dictionary are unordered and, as such, can be displayed in any order (which is arbitrarily decided by Python). Thus, instead of writing your test as

```
"""
>>> myfunction(x)
{'a': 1, 'b': 2, 'c': 3}
"""
```

you should assign the dictionary to a variable and then test whether the variable is == (i.e., equivalent) to the desired answer:

```
"""
>>> tmp = myfunction(x)
>>> tmp == {'a': 1, 'b': 2, 'c': 3}
True
"""
```

In this way, the test will fail only if the dictionaries are not the same, up to a reordering of the keys. Similar considerations hold for functions returning sets.

The Function Depends on a Random Value

Suppose that the function depends on a random value. Then each time you are going to find a different answer. How do we test this? For example, take this function that was saved as testrnd.py:

```
import scipy

def get_sample_mean(n):
    my_sample = scipy.random.normal(size = n)
    return scipy.mean(my_sample)
```

Running it a few times, you will obtain different results:

```
In [2]: get_sample_mean(10)
Out[2]: -0.47708389944942609
In [3]: get_sample_mean(10)
Out[3]: 0.39335892922537397
```

However, you can set the random number generator to repeat a certain random sequence of numbers by setting its *seed*:

```
In [4]: scipy.random.seed(123)
In [5]: get_sample_mean(10)
Out[5]: -0.26951611032632794
In [6]: scipy.random.seed(123)
In [7]: get_sample_mean(10)
Out[7]: -0.26951611032632794
```

Therefore, a possible test could look like

```
import scipy

def get_sample_mean(n):
    """ For testing, we want to make sure we
        set the seed of the random number generator:

        >>> scipy.random.seed(1)
        >>> get_sample_mean(10) # doctest: +ELLIPSIS
        -0.0971...
    """
    my_sample = scipy.random.normal(size = n)
    return scipy.mean(my_sample)
```

doctest is a simple and yet powerful tool for writing tests. If you want to write more complex tests, you can check out one of the many (more sophisticated) modules available in Python: unittest, nose, and mock, among others. One benefit of using doctest for simple programs is that it forces you to maintain up-to-date documentation meaning that the tests can be read as a manual on how to call a function, and what the output should look like.

4.8 Profiling

As Donald Knuth put it, "Premature optimization is the root of all evil," meaning that in most cases, trying to optimize the code for speed is a major source of errors (and, paradoxically, a time sink). This is especially true for science, where we care first and foremost that the code is correct, and only then do we consider speed of execution.

That said, when speed is indeed important, we want to find out where the program is spending most of its time and tweak only those components of our code. This operation is called *profiling* the code and Python makes it easy. Within Jupyter, invoke the program you want to profile (e.g., `simulate_drift`) with this special command:

```
# call simulate_drift.py with N = 1000 and p = 0.1
In [1]: %run -p simulate_drift.py 1000 0.1
```

where the `-p` option of `%run` invokes the Python profiler.

On the computer that we used to write this book, we receive the following output:

```
        10327728 function calls in 3.154 seconds

  Ordered by: internal time

  ncalls  tottime  percall  cumtime  percall filename:
       ↳ lineno(function)
    2576    2.172    0.001    2.892    0.001 drift.py:35(
       ↳ reproduce_population)
 7728000    0.569    0.000    0.569    0.000 {method '
     ↳ randint' of 'mtrand.RandomState' objects}
   10308    0.216    0.000    0.216    0.000 {method '
     ↳ count' of 'list' objects}
 2577008    0.134    0.000    0.134    0.000 {method '
     ↳ append' of 'list' objects}
       1    0.041    0.041    3.154    3.154
          ↳ simulate_drift.py:6(simulate_drift)
  ...
```

You can scroll up and down with your arrow keys. To quit, click on the little "x" in the top-right corner. This output tells us that most of the time is spent on the function reproduce_population, which is called (in this case) 2576 times. This function, in turn, calls the function randint many times (7,728,000 in fact!), so it's not surprising that randint is the second most time-intensive function.

In this case, if we need to speed up the code, we can work on the reproduction function to try to make it faster (without giving up readability). For example, open the file drift.py in a text editor, copy the function reproduce_population to a cell of the Jupyter notebook, and in the next cell write a slightly altered version of the function that we call reproduce_population2:[7]

```python
def reproduce_population2(population):
    """ Create new generation through reproduction
        For each of N new offspring,
        - choose the parents at random;
        - the offspring receives a chromosome from
            each of the parents.
    """
    N = len(population)
    new_generation = [("")] * N
    rai = scipy.random.randint
    for i in range(N):
        # random integer between 0 and N-1
        dad = rai(N)
        mom = rai(N)
        # which chromosome comes from mom
        chr_mom = rai(2)
        offspring = (population[mom][chr_mom], population
            ↳ [dad][1 - chr_mom])
        new_generation[i] = offspring
    return(new_generation)
```

We have made two modifications: (a) we generate the list with all the individuals, and assign new_generation[i] = offspring instead of appending, and (b) we have created a shortcut for the function rai =

---

7. This would be a good time to create a branch in your project's repository as discussed in section 2.6. In this way, you would not clutter your project directory with similar scripts that are difficult to distinguish (especially without proper documentation). You also wouldn't run the risk of losing the functioning, albeit slower, version.

scipy.random.randint, so that Python does not lose time looking for it. These are extremely small modifications: Will they improve the execution time? To compare, we use the command %timeit which executes a function multiple times and returns the execution time. We run it on the old and new versions of our function (i.e., reproduce_population and reproduce_population2):

```
In [4]: %run drift.py
In [5]: mypop = build_population(500, 0.1)
In [6]: %timeit -r 10 reproduce_population(mypop)
10000 loops, best of 10: 64.6 micros per loop
In [7]: %timeit -r 10 reproduce_population2(mypop)
10000 loops, best of 10: 44.7 micros per loop
```

The option -r 10 of the %timeit command specifies that we want to run the code 10 times and return the best timing. As we can see, the minute changes we introduced went a long way to speeding up our simulation.

To reiterate, clarity and correctness come first, while speed is only a secondary concern. If speed becomes an issue, use a profiler to find out where your code is spending its time, make small changes (running your tests after each change to make sure the functions are still behaving correctly), and act only upon the bottlenecks.

4.9 Beyond the Basics

In this final section, we examine a few more advanced topics, which help us to understand the mechanics of Python. We show how common arithmetic operators have been repurposed to work with data structures, detail the differences between mutable and immutable types, explain how to properly copy objects, and determine which variables are visible from within a function.

4.9.1 Arithmetic of Data Structures

In Python, the usual arithmetic operators have been repurposed for working with data structures as well. For example, try concatenating objects using +:

```
In [1]: a = [1, 2, 3]
In [2]: b = [4, 5]
In [3]: a + b
```

```
Out[3]: [1, 2, 3, 4, 5]
In [4]: a = (1, 2)
In [5]: b = (4, 6)
In [6]: a + b
Out[6]: (1, 2, 4, 6)
In [7]: z1 = {1: "AAA", 2: "BBB"}
In [8]: z2 = {3: "CCC", 4: "DDD"}
In [9]: z1 + z2
------------------------------------------------

TypeError  Traceback (most recent call last)
----> 1 z1 + z2
TypeError: unsupported operand type(s) for +: 'dict' and '
    ↳ dict'
# we cannot add two dictionaries with + (use .update!)
```

Similarly, try using * on strings, lists, and tuples:

```
In [10]: "a" * 3
Out[10]: 'aaa'
In [11]: (1, 2, 4) * 2
Out[11]: (1, 2, 4, 1, 2, 4)
In [12]: [1, 3, 5] * 4
Out[12]: [1, 3, 5, 1, 3, 5, 1, 3, 5, 1, 3, 5]
```

4.9.2 Mutable and Immutable Types

When defining variables, Python distinguishes between *mutable* and *immutable* types. As the name implies, we can modify variables that contain *mutable* types, while we cannot modify *immutable* ones. Typically, data structures functioning as "containers" are mutable. For example, if we have a list, we can modify its contents:

```
In [1]: a = [1, 2, 3]
In [2]: a[0] = 1000
In [3]: a
Out[3]: [1000, 2, 3]
```

Similarly, we can update a dictionary by adding key:value pairs or by modifying the values:

```
In [4]: dd = {"a": 1}
In [5]: dd.update({"b": 2}) # this changes the dictionary
In [6]: dd
Out[6]: {'a': 1, 'b': 2}
In [7]: dd["a"] = dd.get("a", 0) + 1
In [8]: dd
Out[8]: {'a': 2, 'b': 2}
```

A tuple, on the other hand, is *immutable*: we cannot modify it, but only substitute it completely (in fact, there is no append method for tuples). Strings, floats, and integers are also immutable. We cannot update their values, only substitute them.

The difference between mutable and immutable objects is important in certain situations. For example, when we are operating on a mutable type, changes are typically happening in place:

```
In [9]: a
Out[9]: [1000, 2, 3]
In [10]: a.sort()   # sort modifies a list in place
In [11]: a  # "a" has changed
Out[11]: [2, 3, 1000]
In [12]: b = a.pop()
In [13]: a
Out[13]: [2, 3]
```

On the other hand, when operating on immutable types, such as tuples and strings, we are not altering the object in any way, but simply returning a different object:

```
In [14]: tt = (1, 2, 3) # create tuple "tt"
In [15]: tt + (4, 5) # this creates a new tuple
Out[15]: (1, 2, 3, 4, 5)
In [16]: tt # the original is unchanged
Out[16]: (1, 2, 3)
```

```
In [17]: ss = "a string"
In [18]: ss.upper() # returns a new string
Out[18]: 'A STRING'
In [19]: ss # the original is unchanged
Out[19]: 'a string'
```

4.9.3 Copying Objects

When trying to copy objects in Python, one might encounter problems, typically stemming from using the assignment function (=) to copy a variable. Assigning and copying are distinct processes. Here we show common mistakes, and their solutions.

If you try to copy an immutable object, such as a number or a string, using assignment, everything is good:

```
In [1]: a = 15
In [2]: b = a
In [3]: a = 32
In [4]: a
Out[4]: 32
In [5]: b
Out[5]: 15
```

When dealing with more complex data structures, however, the assignment a = b does *not* copy the object:

```
In [6]: a = [1, 2, 3]
In [7]: b = a
In [8]: a.append(4)
In [9]: a
Out[9]: [1, 2, 3, 4]
# we append to a after we (attempt to) copy it to b;
# however, b has also been extended!
In [10]: b
Out[10]: [1, 2, 3, 4]
```

When you want to copy lists or dictionaries, you should use their copy function:

```
In [11]: a = [1, 2, 3]
In [12]: b = a.copy()
In [13]: a.append(4)
In [14]: a
Out[14]: [1, 2, 3, 4]
In [15]: b
Out[15]: [1, 2, 3]
```

This will work well if the list or dictionary has only one level. For nested lists, or more complex objects, you can create a copy using the function deepcopy of the module copy:

```
# create list of lists
In [16]: a = [[1, 2], [3, 4]]
# assign copy of a to b
In [17]: b = a.copy()
# modify second value of first list of a
In [18]: a[0][1] = 10
In [19]: a
Out[19]: [[1, 10], [3, 4]]
# b has changed, too!
In [20]: b
Out[20]: [[1, 10], [3, 4]]
In [21]: import copy
In [22]: a = [[1, 2], [3, 4]]
# assign deep copy of a to b
In [23]: b = copy.deepcopy(a)
In [24]: a[0][1] = 10
In [25]: a
Out[25]: [[1, 10], [3, 4]]
# b has not changed
In [26]: b
Out[26]: [[1, 2], [3, 4]]
```

Practically speaking, simply use copy or deepcopy whenever you need to duplicate an object. Note, however, that this is not a "bug" in Python, but rather a direct consequence of how Python stores the objects in the memory of the computer.[8]

4.9.4 Variable Scope

The scope of a variable name describes in which part of the code the variable can be accessed. There are two main types of scope you want to be aware of: *global scope* and *function scope*. Let's start with an example that illustrates the difference. In a cell, type

```
def changea(x):
    a = x
    print("New value for a:", a)


a = 51
print("Current value of a:", a)
changea(22)
print("After calling the function:", a)
```

and run it. You should see

```
Current value of a: 51
New value for a: 22
After calling the function: 51
```

Why didn't calling the function changea affect the value of a? This is because the variable a within the function is not the same as the variable a outside the function!

This is an important feature of all programming languages: you can use any name for the variables within your functions, without the need to constantly make sure they are all distinct. The program will automatically "mask" the names for you: you can think of it as Python using the name of the function as the "last name" of each variable, so that you can give variables in different functions the same "first name" without causing confusion. This is very handy, especially when building large programs, drawing on a number of functions.

---

8. You can learn more about the way Python manages assignment and copy operations in this detailed, yet nontechnical, explanation: computingskillsforbiologists.com/deepcopy.

Sometimes, however, you might need to link the name of a variable within a function (*function scope*) with a variable of the same name outside the function (*global scope*). This can be accomplished by using the statement global immediately after the definition of the function:

```
def changea(x):
    global a
    a = x
    print("New value for a:", a)


a = 51
print("Current value of a:", a)
changea(22)
print("After calling the function:", a)
```

which should return

```
Current value of a: 51
New value for a: 22
After calling the function: 22
```

4.10 Exercises

4.10.1 Assortative Mating in Animals

Jiang et al. (2013) studied assortative mating in animals. They compiled a large database, reporting the results of many experiments on mating. In particular, for several taxa they provide the value of correlation among the sizes of the mates. A positive value of r stands for assortative mating (large animals tend to mate with large animals), and a negative value for disassortative mating.

1. You can find the data in good_code/data/Jiang2013_data.csv. Write a function that takes as input the desired Taxon and returns the mean value of r.
2. You should see that fish have a positive value of r, but that this is also true for other taxa. Is the mean value of r especially high for fish? To test this, compute a *p-value* by repeatedly sampling 37 values of r (37 experiments on fish are reported in the database) at random, and calculating

the probability of observing a higher mean value of r. To get an accurate estimate of the *p-value*, use 50,000 randomizations.

3. Repeat the procedure for all taxa.

4.10.2 Human Intestinal Ecosystems

Lahti et al. (2014) studied the microbial communities living in the intestines of 1000 individuals. They found that bacterial strains tend to be either absent or abundant, and posit that this would reflect bistability in these bacterial assemblages. The data used in this study are contained in the directory good_code/data/Lahti2014. The directory contains the file Metadata.tab characterizing each of the 1006 human records, the file HITChip.tab containing HITChip signal estimates of microbial abundance, and README, a description of the data by the study authors.

1. Write a function that takes as input a dictionary of constraints (i.e., selecting a specific group of records) and returns a dictionary tabulating the BMI group for all the records matching the constraints. For example, calling

   ```
   get_BMI_count({"Age": "28", "Sex": "female"})
   ```

   should return

   ```
   {'NA': 3, 'lean': 8, 'overweight': 2, 'underweight':
      ↳ 1}
   ```

2. Write a function that takes as input the constraints (as above) and a bacterial "genus." The function returns the average abundance (in logarithm base 10) of the genus for each BMI group in the subpopulation. For example, calling

   ```
   get_abundance_by_BMI({"Time": "0",
                         "Nationality": "US"},
                         "Clostridium difficile et rel.")
   ```

should return

```
-----------------------------------------------
Abundance of Clostridium difficile et rel.
In subpopulation:
-----------------------------------------------
Nationality -> US
Time -> 0
-----------------------------------------------
3.08    NA
3.31    underweight
3.84    lean
2.89    overweight
3.31    obese
3.45    severeobese
-----------------------------------------------
```

3. Repeat this analysis for all genera, and for the records having Time = 0.

4.11 References and Reading

Style Guides

Python's PEP-8 document describing writing style:
computingskillsforbiologists.com/pythondocstyleguide.

Google's style guide for Python:
computingskillsforbiologists.com/pythonstyleguide.

Pylint, a program to automatically check that your code is well written:
pylint.org.

Integrated Development Environment

Jupyter
The documentation is available at
jupyter.readthedocs.io/.

Unit Testing

The doctest documentation:
computingskillsforbiologists.com/doctest.

Introduction to nose and other modules for unit testing:
pythontesting.net.

Magic Commands

We've used some of IPython's "magic commands": %run, %pdb, %timeit. These are short commands that perform complex operations, which would require much code otherwise. Many more are available, each more magic than the last:
computingskillsforbiologists.com/pythonmagic.

Zen

When programming, you constantly have to make decisions. Should I rewrite this piece of code? Should I use this or that data structure? You can find solace and inspiration in the Zen of Python.

```
In [1]: import this
```

CHAPTER 5

•••••••••••

Regular Expressions

5.1 What Are Regular Expressions?

Sometimes, you need to extract data from text. For example, you might want to extract all the protein accession numbers from a paper, the DNA motifs from sequence data, or the geographical coordinates of your sample sites from a large and complicated text file. Often, it is not feasible to search for all possible occurrences exactly as they appear in the text, but you can describe the pattern you're looking for in your own words (e.g., find all words starting with 3 uppercase letters, followed by 4 digits). The question is how to explain such a pattern to a computer. The answer is to use regular expressions.

Regular expressions are used to find a match for a particular pattern in a string of text. We've already used them in section 1.6.5: the Unix command `grep` stands for "global regular expression print." There, we conducted exclusively *literal* searches, meaning that we searched for lines containing an exact match of the input we provided. The power of regular expressions, however, is that we can use special syntax to describe patterns in a general way (e.g., find anything that looks like a Latin binomial), and then easily list all the occurrences of a pattern in a string or text file.

5.2 Why Use Regular Expressions?

Ask several programmers what they think about regular expressions, and you might hear that they are the greatest thing since sliced bread, or one of the greatest nuisances ever invented. Despite the polarized opinion, for many biological problems, regular expressions can save the day. They can be used

to collect information: Search for patterns corresponding to structural or functional features in sequence data (e.g., degenerated primer binding sites, transcription factor binding sites). Similarly, simple searches can match accession and gene numbers, or extract references from a manuscript.

to navigate and parse text files: Whenever the text is semistructured (e.g., the output produced by a machine), it is easy to digest it using regular expressions. There is no need to use regular expressions to parse well-structured HTML, XML, or JSON files—there are specialized tools for that!—but for other, idiosyncratic formats, regular expressions can get the job done.

as a more sophisticated replace: Sometimes, you want to replace all strings that are similar to a given string (e.g., with variations in capitalization or spaces). Regular expressions allow you to find and replace strings according to patterns.

to shorten your code: A few lines of regular expressions can replace many lines of code. This is a controversial aspect since regular expressions are compact, but notoriously difficult to read. In section 5.7 we introduce best practices to alleviate this problem.

Regular expressions are quite ubiquitous. We have already used grep to search a text file in your shell. Most dedicated text editors for programming (such as gedit, emacs, and vim) allow you to search using regular expressions. Even better, almost every programming language either implements regular expressions natively (e.g., Perl, JavaScript, Ruby) or provides special packages implementing them (e.g., Java, Python, C++). The downside of this popularity is that each programming language implements a specific "dialect" of the syntax for regular expressions, with slight (yet annoying) variations. In the following, we explore regular expressions using Python.

5.3 Regular Expressions in Python

First, we introduce the module re which implements regular expressions in Python. With that at hand, we learn how to construct regular expressions, working on simple examples. Finally, we put it all together, showing how powerful regular expressions can be, and working through more complex examples to validate, manipulate, and extract data from text.

5.3.1 The re Module in Python

The functions dealing with regular expressions in Python are contained in the module re. Thus, we start our Jupyter session with

```
In [1]: import re
```

The simplest function of the re module is re.search. This function requires two arguments: the pattern you want to match (i.e., the regular expression) and the target string that you want to search. The function finds *only* the first match (if a match is found). It returns a *match object* if the pattern is found, and None if the pattern is not found. For example, type

```
In [2]: my_string = "a given string"
In [3]: m = re.search(r"given", my_string)
```

Printing the *match object* (that we have assigned to m) returns a slightly puzzling message:

```
In [4]: print(m)
<_sre.SRE_Match object; span=(2, 7), match='given'>
```

If a match is found, the *match object* provides information on the location of the match within the target string (span) and the identified match (match). Usually, we do not print the match object but extract the identified match directly by calling the group method:

```
In [5]: m.group()
Out[5]: 'given'
```

If nothing is matched, the function re.search returns None:

```
In [6]: m = re.search(r"9", my_string)
In [7]: print(m)
None
```

5.4 Building Regular Expressions

Now that we are familiar with the simplest functions in the re module, we can practice building regular expressions. Note that the syntax of regular expressions in this section is common to all programming languages. While we use Python for our examples, you could easily use them in another programming language, or in your text editor, to define a search.

5.4.1 Literal Characters

A regular expression defines a pattern describing what we want to find. The simplest form of a regular expression is a *literal* character or string, which means that we are trying to match exactly the characters we typed. We did that in the previous section, where we searched literally for the word given. This is easy to do, but has limited use—often it is not feasible to search for each and every string that we might want to match.

5.4.2 Metacharacters

The real power of regular expressions comes with using *metacharacters*. Metacharacters stand in for a well-defined class of other characters. We define metacharacters by prepending a backslash (\). The backslash functions as an *escape character* and signals the regular expression parser that what follows needs to be "interpreted" as a metacharacter (e.g., \n matches a newline character, rather than the literal character n). The following box lists some examples of metacharacters and their meaning.

Metacharacter	Meaning
\n	Match a newline.
\t	Match a tab.
\s	Match white space (i.e., tab, space, newline, or carriage return).
\w	Match a "word character" (alphanumeric & underscore).
\d	Match a digit.
.	Match any character.

The metacharacter \s stands for white space (i.e., tab, space, newline, or carriage return):

```
In [8]: my_string = "a given string"
# match white space
In [9]: m = re.search(r"\s", my_string)
In [10]: m.group()
Out[10]: ' '
```

Now we match a white space followed by five "word" characters (alphanu-meric and underscore):

```
# 1 white space followed by 5 "word" characters
In [11]: m = re.search(r"\s\w\w\w\w\w", my_string)
In [12]: m.group()
Out[12]: ' given'
```

Note that this could also have matched " strin", but re.search returns only the first instance.

5.4.3 Sets

Sets, defined by square brackets, are used to provide several options or a range of characters to match. Let's look at an example:

```
In [13]: my_string = "sunflowers are described on page 89"
# search for word that starts with lowercase or uppercase
# letter s followed by two word characters
In [14]: m = re.search(r"[sS]\w\w", my_string)
In [15]: m.group()
Out[15]: 'sun'
# search for a number with two digits
In [16]: m = re.search(r"[0-9][0-9]", my_string)
In [17]: m.group()
Out[17]: '89'
```

Maybe you have realized that the set [0-9] is equivalent to using the single metacharacter \d. Likewise, the metacharacter \w defines the set [a-zA-Z0-9_]. In comparison to metacharacters, sets allow you to be even more specific about what should be matched at a certain position in your target string.

Using a caret (^) as the first symbol within a set negates the set (i.e., match any character not listed in the set):

```
# match a character not in s-z
# followed by 6 word characters
```

```
In [18]: m = re.search(r"[^s-z]\w\w\w\w\w\w", my_string)
In [19]: m.group()
Out[19]: 'nflower'
```

5.4.4 Quantifiers

In our previous search, we typed the same metacharacter (\w) several times. Instead, we can use *quantifiers*, listed below, to determine how many characters of a certain type to expect. Our search to match a white space followed by exactly five word characters could therefore also be typed

```
# 1 white space followed by exactly 5 "word" characters
In [20]: m = re.search(r"\s\w{5}", my_string)
In [21]: m.group()
Out[21]: ' descr'
```

Quantifier	Meaning
?	Match zero or one time.
*	Match zero or more times.
+	Match one or more times.
{n}	Match exactly n times.
{n,}	Match at least n times.
{n,m}	Match at least n but not more than m times.

As an example, we attempt to extract a DNA sequence from text:

```
# 1 or more capital letters A, C, G, or T
In [22]: re.search(r"[ACGT]+","A possible explanation is
    ↳ the motif ATTCGT.").group()
Out[22]: 'A'
```

Our regular expression describes the pattern "one or more capital letters A, C, G, or T." The beginning of our target string already fulfills this pattern, and therefore we have found an unintended match. We have to write a more specific regular expression in order to match only the nucleotide sequence:

```
In [23]: re.search(r"[ACGT]{3,}", "the motif ATTCGT").
    ↳ group()
Out[23]: 'ATTCGT'
```

Here, we have specified that we want to return a match only if there are at least three consecutive capital letters A, C, G, or T, which in turn matches only our DNA sequence.

Note that the quantifiers ?, *, and + try to match as much text as possible—they are *greedy*:

```
In [24]: my_string = "once upon a time"
# match any number of characters
# followed by a white space
In [25]: re.search(r".*\s", my_string)
In [26]: m.group()
Out[26]: 'once upon a '
```

The word "once " alone would satisfy the regular expression, but the quantifier (*) makes the expression greedy. Appending a question mark to the quantifier (i.e., ??, *?, or +?) makes it *reluctant* or *nongreedy* so that it will match as little as possible instead:

```
In [27]: re.search(r".*?\s", "once upon a time").group()
Out[27]: 'once '
```

5.4.5 Anchors

Sometimes, we might want to *anchor* our regular expressions. This means that we expect our pattern to be either at the end or at the beginning of the string. Use the caret symbol (^) to match the beginning of a string, and the dollar sign ($) to match the end:

```
# the pattern is present, but not at the beginning of the
# string: the function does not find a match
In [28]: my_string = "ATATA"
In [29]: m = re.search(r"^TATA", my_string)
```

```
In [30]: print(m)
None
# searching at the end of the string is successful
In [31]: m = re.search(r"TATA$", my_string)
In [32]: m.group()
Out[32]: 'TATA'
```

Intermezzo 5.1

Describe the following regular expressions in plain English. What does the regular expression match? You can type each command into your notebook to see the result.

(a) re.search(r"\d" , "it takes 2 to tango").group()
(b) re.search(r"\w*\s\d.*\d", "take 2 grams of H20").group()
(c) re.search(r"\s\w*\s", "once upon a time").group()
(d) re.search(r"\s\w{1,3}\s", "once upon a time").group()
(e) re.search(r"\s\w*$", "once upon a time").group()

5.4.6 Alternations

Last but not least, we can use the pipe symbol ($|$) to provide *alternations*. The symbol tells the regular expression parser to match either everything to the left of the pipe symbol, or everything to the right of it.

```
# match
In [33]: my_string = "I found my cat!"
In [34]: m = re.search(r"cat|mouse", my_string)
In [35]: m.group()
Out[35]: 'cat'
```

With a mixture of literals and metacharacters, we can now describe any pattern.

Intermezzo 5.2

Let's practice translating from plain English to regular expressions. The NCBI GenBank contains information on nucleotide sequences, protein sequences, and whole genome sequences (WGS).[a] The following table describes the construction of sequence identifiers in plain English. Construct the appropriate regular expression to match either protein, WGS or nucleotide ids.

(a) Protein: 3 letters + 5 numerals
(b) WGS: 4 letters + 2 numerals for WGS assembly version + 6–8 numerals
(c) Nucleotide: 1 letter + 5 numerals OR 2 letters + 6 numerals

---

a. https://www.ncbi.nlm.nih.gov/Sequin/acc.html.

5.4.7 Raw String Notation and Escaping Metacharacters

You might have noticed that we put an r in front of our regular expressions. This means that we want to evaluate the string in its "raw" form. Without it, Python would automatically interpret any special character (i.e., metacharacter). For example, it would insert a newline for any \n. Here, we want to prevent such interpretation and use the *raw* string. Let's take a look at the difference between raw and regular strings by printing a statement:

```
# Python interprets the metacharacter
In [1]: print("my long \n line")
my long
 line
# use of the raw format of the string (by prepending an r)
# means that Python will not interpret the metacharacter
In [2]: print(r"my long \n line")
my long \n line
```

So what shall we do when we actually want to search for something that looks like a metacharacter, such as \n, in a raw string? In this case, we have to escape the metacharacter with another "\", as in the following example:

```
In [3]: my_string = r"my long \n line"
In [4]: m = re.search(r"\\n", my_string)
In [5]: m.group()
Out[5]: '\\n'
```

Let's look at a more comprehensive example. We want to find the names of variants of MHC alleles in the file CSB/regex/data/Marra2014_ BLAST_data.txt. The file is the result of a BLAST search using the next generation sequencing contigs that we have previously dealt with in chapter 1 and exercise 1.10.1. The MHC variants contain a * in their name. If we want to search literally for the asterisk symbol, we have to escape it using the backslash (otherwise it will be interpreted as a metacharacter):

```
# open file and generate a file handle
In [6]: with open("../data/Marra2014_BLAST_data.txt")
    ↳ as f:
    # initiate counter for matches
    counter = 0
    # search for pattern in each line
    for line in f:
        m = re.search(r"\*", line)
    # only if a match was found, increase counter
    # and print the line
        if m:
            counter += 1
            print(line)
# print information on how many lines contain a match
print("The pattern was matched in {0} lines".format
    ↳ (counter))

contig01987 2b14_human ame: full=hla class ii
    ↳ histocompatibility drb1-4
[...]
The pattern was matched in 6 lines
```

The code prints all the lines that contain an asterisk and informs us that 6 lines contain a match. We also see that the last 2 lines contain an asterisk but are not MHC alleles. Now that we have an idea what our matches look like, we can refine our regular expression to exclude unwanted lines, returning only the full name of the MHC allele:

```
In [7]: with open("../data/Marra2014_BLAST_data.txt")
    ↳ as f:
    for line in f:
```

```
        m = re.search(r"mhc[\s\w*]+\*\w*", line)
        if m:
            print(m.group())
```

```
mhc class ii antigen drb1*4
mhc class i antigen cw*8
mhc class i antigen b*46
mhc class ii antigen drb1*1
```

These are the individual parts of the regular expression:

mhc	Match the literal characters mhc.
[\s\w*]+	Match a white space or zero or more word characters, one or more times (the + makes it greedy).
*	Match a (literal) asterisk.
\w*	Match zero or more word characters.

5.5 Functions of the re Module

So far, we have exclusively worked with the function re.search. However, the module re provides many more functions. Find an overview of the most commonly used ones in the box below. Usage examples follow in subsequent sections.

---

re.findall(reg, target_string) As re.search, but return a list of *all* the matches.

re.finditer(reg, target_string) As re.search, but return an iterator, that is, an object you can call to access the next match, as in

```
for my_match in re.finditer(reg, target_string):
    print(my_match.group())
```

re.compile(reg) Compile a regular expression. In this way, the pattern is stored for repeated use, improving the speed.

re.split(reg, target_string) Split the text by the occurrence of the pattern described by the regular expression.

re.sub(reg, repl, target_string) Substitute each nonoverlapping occurrence of the match with the text in repl.

As always, in Jupyter you can type re. and press Tab to show a list with all functions in the re module. The help also provides information on functions, as well as an extended list of metacharacters. To access the help for the re module, type

```
help(re)
```

In the following, we explore several functions of the re module. We work with re and the pyfaidx module,[1] which facilitates working with FASTA files. We want to find methylation sites in nucleotide sequences from *Escherichia coli* that are stored in the file CSB/regex/data/Ecoli.fasta. We first identify the motif GATC, which is the target site of the most common methylase, called Dam. Go to regex/sandbox and launch a new Jupyter notebook.

```
In [1]: import re
In [2]: import pyfaidx
# read the .fasta file as pyfaidx.Fasta data type
In [3]: genes = pyfaidx.Fasta("../data/Ecoli.fasta")
```

The pyfaidx.fasta object behaves like a dictionary. The keys are the names of the FASTA sequences. These are very long in our case, which makes it difficult to call the dictionary elements by name. We could either shorten the FASTA identifier (but lose information), or use the following work-around: we create a list with the dictionary keys (i.e., names of FASTA records). We can then call individual sequences by indexing our list with record names:

```
# convert keys of dictionary to list
In [4]: records = list(genes.keys())
In [5]: records
Out[5]: ['gi|556503834|ref|NC_000913.3|:1978338-2028069',
         'gi|556503834|ref|NC_000913.3|:4035299-4037302']
# extract first sequence from genes dictionary
# get sequence start to finish
In [6]: seq1 = genes[records[0]][:]
```

---

1. For installation of the pyfaidx module, see computingskillsforbiologists.com/pyfaidx.

```
# explore pyfaidx.Sequence object
seq1.    # hit Tab
# call end attribute to determine length of sequence
In [7]: seq1.end
Out[7]: 49732
# the seq function returns our sequence as a string,
# a prerequisite to applying re functions
seq1_str = seq1.seq
# print the first 40 nucleotides
In [8] seq1_str[:40]
Out[8]: 'AATATGTCCTTACAAATAGAAATGGGTCTTTACACTTATC'
```

Now that we have the target sequence in place, we can search for our pattern and retrieve information on the position of the match:

```
# literal search for the pattern "GATC"
In [9]: m = re.search(r"GATC", seq1_str)
In [10]: m.group()
Out[10]: 'GATC'
# extract the start and end positions of the match
# remember: Python starts counting at 0
In [11}: m.start()
Out[11]: 130
# noninclusive, similar to range function
In [12]: m.end()
Out[12]: 134
```

There are actually several methylases in *E. coli*. The EcoKI methylase, for example, modifies the sequences AACNNNNNNGTGC and GCAC-NNNNNNGTT (where "N" stands for any nucleotide). However, the function re.search returns information for only the first pattern that was found, so we would miss other matches in our sequence. If we want to find all possible methylation sites in our sequence, we have to use the function re.findall:

```
In [13]: m = re.findall(r"AAC[ATCG]{6}GTGC|GCAC[ATCG]{6}
    ↳ GTT", seq1_str)
# the pattern says AAC followed by 6 of A,T,C, or G,
# followed by GTGC; or GCAC followed by 6 of A,T,C, or G
```

```
# followed by GTT;
# test whether match was found and show formatted results
In [14]: if m:
             print("There are " + str(len(m)) + " matches"
                ↳ )
             print("These are the matches: " + str(m))
There are 6 matches
These are the matches: ['AACAGCATCGTGC', 'AACTGGCGGGTGC',
    ↳ ...]
```

Note that re.findall returns a list and not a match object (i.e., we can apply the function len but not the method .group or find the position of the match object via method .start). In order to retrieve positions for all matches, we need the function re.finditer, which returns an iterator that we can use in a loop:

```
In [15]: hits = re.finditer(r"AAC[ATCG]{6}GTGC|GCAC[ATCG
    ↳ ]{6}GTT", seq1_str)
In [16]: for item in hits:
                print(item.start() + 1, item.group())

18452 AACAGCATCGTGC
18750 AACTGGCGGGTGC
25767 GCACCACCGCGTT
35183 GCACAACAAGGTT
40745 GCACCGCTGGGTT
42032 AACCTGCCGGTGC
```

It is important to remember that Python starts counting at zero. We take this into account by adding (+1) to the start of the match when we print the position of the match in the sequence.

The regular expression to find the EcoKI sites is more complex and tedious to type. If we want to use it many times in our program, we can decrease programming and run times by compiling the regular expression, which turns it into a Python object that inherits all re functions:

```
In [17]: EcoKI = re.compile(r"AAC[ATCG]{6}GTGC|GCAC[ATCG
    ↳ ]{6}GTT")
In [18]: m = EcoKI.findall(seq1_str)
```

```
In [19]: m
Out[19]:
['AACAGCATCGTGC',
 'AACTGGCGGGTGC', ...]
```

5.6 Groups in Regular Expressions

Groups are defined by surrounding part(s) of the regular expression with parentheses. Groups can help structure the regular expression or capture substrings of the returned result. Let's look at an example showing how the use of groups alters the match:

```
# match a G, followed by 2 T's
In [1]: re.search(r"GT{2}", "ATGGTGTCCGTGTT").group()
Out[1]: 'GTT'
# match GT twice
In [2]: re.search(r"(GT){2}", "ATGGTGTCCGTGTT").group()
Out[2]: 'GTGT'
```

Furthermore, groups are useful when we need more control over the match than simply returning the matched string. For instance, we might not be able to describe a pattern for the region we want to match, but can do so for the text that flanks our region of interest. In such a case, we use groups to separate the result from the patterns that we used to identify the match (but are actually not interested in).

Here is an example to illustrate this idea: Many taxonomic studies determine species identity by using conserved primers that flank a taxonomically informative (i.e., polymorphic) sequence. The primers can easily be described, but the region of interest is very variable so that matching it is more difficult. In our example, we want to identify the variable region that is flanked by the bacterial ribosomal RNA primers 799F and 904R. Defining our regular expressions in groups facilitates further analysis by separating the informative, variable sequence from our primer sequences.

We work with the second sequence of the file regex/data/Ecoli.fasta. To spare you some typing, we have stored the regular expression pattern for the 16S primer in the file ../data/16S_regex_pattern.txt. Let's first read the file and look at the regex pattern:

```
In [1]: with open("../data/16S_regex_pattern.txt") as f:
            # read first line
            regpat = f.readline()
In [2]: print(regpat)
(AAC[AC]GGATTAGATACCC[GT]G)([ATCG]+)([CT]T[AG]
    ↳ AAACTCAAATGAATTGACGGGG)
```

Note how the pattern is structured with groups of round brackets. The first pair of brackets encloses our forward primer 799F. There is a polymorphic site at the fourth position and another one at the second to last position which are indicated with a set (using square brackets). The second pair of round brackets encloses our variable middle part—what we are actually interested in. All we assume is that the sequence should be more than one nucleotide (i.e., A, T, C, or G) long. The third pair of round brackets encloses our reverse primer 904R, which has polymorphic sites at positions 1 and 3.

We are ready to retrieve the second sequence of our Ecoli.fasta file and search for the pattern.

```
In [1]: seq2_str = genes[records[1]][:].seq
# compile regex pattern
In [2]: amp = re.compile(regpat)
In [3]: m = amp.search(seq2_str)
# now we can individually return the matched groups
# group(0) returns the entire match
# we look at only the first 40 nucleotides
In [4]: m.group(0)[:40]
Out[4]: 'AACAGGATTAGATACCCTGGTAGTCCACGCCGTAAACGAT'
# group(1) returns the first group - our forward primer
In [5]: m.group(1)
Out[5]: 'AACAGGATTAGATACCCTG'
# group(2) is the variable middle part
# this is the sequence we're interested in
In [6]: m.group(2)[:40]
Out[6]: 'GTAGTCCACGCCGTAAACGATGTCGACTTGGAGGTTGTGC'
# we can perform further analysis on this group
In [7]: len(m.group(2))
Out[7]: 104
# group(3) is our reverse primer
In [8]: m.group(3)
Out[8]: 'TTAAAACTCAAATGAATTGACGGGG'
```

5.7 Verbose Regular Expressions

The main downside of regular expressions is that they can become very difficult to read, making them hard to understand and tweak. Even seemingly simple tasks, such as matching a zip code, can yield complex regular expressions. One way to improve readability is to write "verbose" regular expressions, so that we can document what we're doing. For example,

```
# pattern to match a zip code
pattern = r"""
        ^ # start of the string
        (\d{5}) # 5 digits in a group, e.g., 60637
        ([\s-]{1} # optional part starts with white space
            ↳ or -
        \d{4})? # followed by 4 digits, e.g., 60637-1503
        $ # end of string
        """
```

Verbose regular expressions begin and end with """ (i.e., three double quotes). You can use indentation and comments, making the expression much easier to understand, edit, debug; this is also useful when you want to recycle part of it. When you want to use the pattern, you need to add the option re.VERBOSE when calling the function (e.g., re.search):

```
In [1]: re.search(pattern, "60637", re.VERBOSE).group()
Out[1]: '60637'
In [2]: re.search(pattern, "60637 1503", re.VERBOSE).group
    ↳ ()
Out[2]: '60637 1503'
In [3]: re.search(pattern, "60637-1503", re.VERBOSE).group
    ↳ ()
Out[3]: '60637-1503'
```

5.8 The Quest for the Perfect Regular Expression

By this point, you should have found out whether you love (or hate!) regular expressions. In either case, we advocate using regular expressions in a "healthy amount": knowing the theory and practice of regular expressions makes you a more efficient scientist. However, sometimes the "perfect" regular expression can be difficult to write, look fairly complicated, or be difficult to debug. In such cases, sometimes it's easier to have a simpler, more permissive regular expression that matches all of the data we want to match, but also some unwanted matches. An additional line of code can then clean the data further.

It is also important to realize when searching for a dedicated package might be better than spending a lot of time trying to identify the one perfect regular expression that fully captures all desired results but also perfectly excludes unwanted matches.

5.9 Exercises

5.9.1 Bee Checklist

Michael Ruggiero of the Integrated Taxonomic Information System (itis.gov) has led the World Bee Checklist project, aiming to collect taxonomic information on all the bee species in the world. In the file `regex/data/bee_list.txt` you can find a list of about 20,000 species, along with their TSN (the identifier in the ITIS database), and a column detailing the authors and year of publication of documents describing the species.

1. What is the name of the author with the most entries in the database? To find out, you'll need to parse the citations in the file. Note that you need to account for different formats. For example, you'll find

```
(Morawitz, 1877)
Cockerell, 1901
(W. F. Kirby, 1900)
Meade-Waldo, 1914
Eardley & Brooks, 1989
Lepeletier & Audinet-Serville, 1828
Michener, LaBerge & Moure, 1955
```

2. Which year of publication is most represented in the database?

5.9.2 A Map of Science

Where does science come from? This question has fascinated researchers for decades, and has even led to the birth of the field of the "science of science," where researchers use the same tools they invented to investigate nature to gain insights into the development of science itself. In this exercise, you will build a "map of *Science*," showing where articles published in *Science* magazine have originated. You will find two files in the directory regex/data/MapOfScience. The first, pubmed_results.txt, is the output of a query to PubMed, listing all the papers published in *Science* in 2015. You will extract the US ZIP codes from this file, and then use the file zipcodes_coordinates.txt to extract the geographic coordinates for each ZIP code.

1. Read the file pubmed_results.txt, and extract all the US ZIP codes.
2. Create the lists zip_code, zip_long, zip_lat, and zip_count, containing the unique ZIP codes, their longitudes, latitudes, and counts (number of occurrences in *Science*), respectively.
3. To visualize the data you've generated, use the following code:[2]

```python
import matplotlib.pyplot as plt
# let plots be produced within the IPython notebook
%matplotlib inline
# create a scatter plot
plt.scatter(zip_long, zip_lat, s = zip_count, c =
    ↳ zip_count)
plt.colorbar()
# only continental US without Alaska
plt.xlim(-125,-65)
plt.ylim(23, 50)
# add a few cities for reference (optional)
ard = dict(arrowstyle="->")
plt.annotate("Los Angeles", xy = (-118.25, 34.05),
    xytext = (-108.25, 34.05), arrowprops = ard)
plt.annotate("Palo Alto", xy = (-122.1381, 37.4292),
    xytext = (-112.1381, 37.4292), arrowprops= ard)
plt.annotate("Cambridge", xy = (-71.1106, 42.3736),
    xytext = (-73.1106, 48.3736), arrowprops= ard)
plt.annotate("Chicago", xy = (-87.6847, 41.8369),
    xytext = (-87.6847, 46.8369), arrowprops= ard)
plt.annotate("Seattle", xy = (-122.33, 47.61),
    xytext = (-116.33, 47.61), arrowprops= ard)
```

---

2. Though not covered in detail in this book, the module Matplotlib provides sophisticated graphics for Python. See matplotlib.org for extensive documentation.

```
plt.annotate("Miami", xy = (-80.21, 25.7753),
    xytext = (-80.21, 30.7753), arrowprops= ard)
# define size of plot
params = plt.gcf()
plSize = params.get_size_inches()
params.set_size_inches( (plSize[0] * 3, plSize[1] *
    ↳ 3) )
# produce the plot
plt.show()
```

5.10 References and Reading

The chapter on regular expressions in *Dive into Python* by Mark Pilgrim:
   computingskillsforbiologists.com/regextutorial.

The *Python Course* by Bernd Klein offers a part on regular expressions. The graphical representation can help to understand the basic principles:
   computingskillsforbiologists.com/retutorial.

A collection of useful regular expressions:
   regxlib.com.

Translate regular expressions into plain English and test your code on example text:
   regex101.com/#python.

Another nice website for testing regular expressions:
   regexr.com.

CHAPTER 6

•••••••••••

Scientific Computing

6.1 Programming for Science

Python is a general-purpose programming language, but the examples in the previous chapters have shown how even the most basic features of Python can be used to solve biological problems. In this chapter, we highlight a few of the packages and modules that the Python community has developed specifically for scientific applications and biology.

The modules NumPy and SciPy allow you to manipulate large numerical data sets. NumPy provides efficient data structures for vectors and matrices, and SciPy is a large library that includes linear algebra routines, as well as numerical algorithms for performing numerical integration, solving differential equations, implementing statistical and mathematical models, and building complex simulations.

The package pandas facilitates the manipulation, analysis, and visualization of data sets. Last but not least, we introduce some functions of the Biopython project, providing support for many bioinformatics file formats, the analysis and visualization of sequences, and retrieval of information from biological databases.

6.1.1 Installing the Packages

Detailed instructions on how to install these packages can be found in the directory CSB/scientific/installation.

6.2 Scientific Programming with NumPy and SciPy

NumPy is the basic Python package for scientific computing. SciPy is a collection of statistical and mathematical functions meant to extend the features of NumPy. The two packages are typically imported in concert.

6.2.1 NumPy Arrays

First, we explore the basic data type provided by NumPy. Import the package:

```
In [1]: import numpy as np
```

The main data structure provided by NumPy is called array, and is meant to contain numerical vectors (one-dimensional arrays), matrices (two-dimensional arrays), and tensors (n-dimensional arrays). For example, the function np.arange() creates a one-dimensional array (similar to list(range(x))):

```
# create an array
In [2]: a = np.arange(9)
In [3]: a
Out[3]: array([0, 1, 2, 3, 4, 5, 6, 7, 8])
```

Contrary to lists, you can multiply all elements by a constant, or add a constant to all the coefficients, etc., using arithmetic operators. For example, the following code illustrates the different behavior of lists and arrays:

```
# create a list
In [4]: al = list(range(9))
In [5]: al
Out[5]: [0, 1, 2, 3, 4, 5, 6, 7, 8]
# multiply list by 2
In [6]: al * 2
Out[6]: [0, 1, 2, 3, 4, 5, 6, 7, 8, 0, 1, 2, 3, 4, 5, 6, 7
    ↳ , 8]
# multiply array by 2
In [7]: a * 2
Out[7]: array([ 0,  2,  4,  6,  8, 10, 12, 14, 16])
# add 2 to elements of array
In [8]: 2 + a
Out[8]: array([ 2,  3,  4,  5,  6,  7,  8,  9, 10])
# trying to add 2 to elements of list
In [9]: 2 + al
```

```
TypeError Traceback (most recent call last)
<ipython-input-fa2132b154fc> in <module>()
----> 1 2 + al
TypeError: unsupported operand type(s) for +: 'int' and '
    ↳ list'
```

Many attributes and functions are associated with arrays. For example, you can access the number of dimensions, the number of coefficients, and the type of data stored in the array:

```
# length along each dimension
In [10]: a.shape
Out[10]: (9,)
# number of dimensions
In [11]: a.ndim
Out[11]: 1
# data type of content
In [12]: a.dtype.name
Out[12]: 'int64'
# number of elements
In [13]: a.size
Out[13]: 9
```

Arrays provide many useful methods to perform arithmetical and statistical operations:

```
In [15]: a.sum()
Out[15]: 36
In [16]: a.mean()
Out[16]: 4.0
In [17]: a.std()
Out[17]: 2.5819888974716112
In [18]: a.min()
Out[18]: 0
In [19]: a.max()
Out[19]: 8
```

Furthermore, the NumPy library provides many functions that can be applied to all elements of the array, for example,

```
# square root
In [20]: np.sqrt(a)
Out[20]: array([ 0.,   1.,   1.41421356,  1.73205081,  2.,
    2.23606798,  2.44948974,  2.64575131,  2.82842712])
# exponentiation
In [21]: np.exp(a)
Out[21]: array([ 1.00000000e+00, 2.71828183e+00,
    7.38905610e+00, 2.00855369e+01, 5.45981500e+01,
    1.48413159e+02, 4.03428793e+02, 1.09663316e+03,
    2.98095799e+03])
```

There are multiple ways of creating arrays. The simplest is to convert a list of numbers into an array:

```
# convert list to one-dimensional array
In [22] a1 = np.array([1, 2, 3, 4])
        print(a1)
        a1.dtype.name
Out[22]: [1 2 3 4]
         'int64'
In [23]: a1 = np.array([1.0, 2.0, 3.0, 4.0])
        print(a1)
        a1.dtype.name
Out[23]: [ 1.   2.   3.   4.]
         'float64'
# convert list of lists to two-dimensional array
In [24]: m = np.array([[1, 2], [3, 4]])
        m
Out[24]: array([[1, 2],
                [3, 4]])
# determine data type
In [25]: m = np.array([[1, 2], [3, 4]], dtype = float)
        print(m)
        m.dtype.name
Out[25]: [[ 1.  2.]
         [ 3.  4.]]
         'float64'
```

Several functions are available to initialize arrays with a given structure:

```
# create 3x2 array filled with 0.0 (floating-point)
In [26]: m = np.zeros((3, 2), dtype = float)
         m
Out[26]: array([[ 0.,  0.],
                [ 0.,  0.],
                [ 0.,  0.]])
# create 2x3 array filled with 1+0i (complex numbers)
In [27]: m = np.ones((2, 3), dtype = complex)
         m
Out[27]: array([[ 1.+0.j,  1.+0.j,  1.+0.j],
                [ 1.+0.j,  1.+0.j,  1.+0.j]])
# create an array containing the integers from 0 to 8
# and arrange it in a 3x3 array
In [28]: a = np.arange(9)
         a.reshape((3,3))
Out[28]: array([[0, 1, 2],
                [3, 4, 5],
                [6, 7, 8]])
# create an array with random values
# drawn from uniform distribution U[0,1]
In [29]: np.random.random((2, 3))
Out[29]: array([[ 0.2331427 ,  0.28167952,  0.66094357],
                [ 0.13703488,  0.75519455,  0.08413554]])
```

You can access the elements using indices, similarly to what you saw for lists. The difference is that now you can take multidimensional slices:

```
In [30]: a
Out[30]: array([0, 1, 2, 3, 4, 5, 6, 7, 8])
# access a coefficient
In [31]: a[1]
Out[31]: 1
# create a slice
In [32]: a[:4]
Out[32]: array([0, 1, 2, 3])
# you can use negative indices
In [33]: a[-3:]
Out[33]: array([6, 7, 8])
```

```
In [34]: m = np.arange(16).reshape((4, 4))
         m
Out[34]: array([[ 0,  1,  2,  3],
                [ 4,  5,  6,  7],
                [ 8,  9, 10, 11],
                [12, 13, 14, 15]])
# extract submatrix
In [35]: m[0:3, 1:4]
Out[35]: array([[ 1,  2,  3],
                [ 5,  6,  7],
                [ 9, 10, 11]])
# this is the second row
In [36]: m[1]
Out[36]: array([4, 5, 6, 7])
# this is the second column
In [37]: m[:, 1]
Out[37]: array([ 1,  5,  9, 13])
```

As you saw above, you can perform operations on the whole array. When you have multidimensional arrays, however, you can operate by row, by column, etc.

```
# whole matrix operation
In [38]: m.sum()
Out[38]: 120
# col sums
In [39]: m.sum(axis = 0)
Out[39]: array([24, 28, 32, 36])
# row sums
In [40]: m.sum(axis = 1)
Out[40]: array([ 6, 22, 38, 54])
```

Finally, it's easy to read data into NumPy: just use the function np. loadtxt,[1] which is very flexible. For example, to read a matrix stored in a text format (such as .csv) into an array, use

```
m = np.loadtxt("my_matrix_file.csv", delimiter = ",")
```

---

1. Documented at computingskillsforbiologists.com/numpyloadtxt.

Example: Image Processing with NumPy

In the following example, we use the NumPy library for image processing. In a computer, an image is typically stored as a three-dimensional numerical array. The height of the image corresponds to the *y*-axis and the width of the image to the *x*-axis.[2] The third dimension refers, for example, to the red, green, or blue channel in an RGB color image.[3] We load two functions from the scikit-image library that are specific to image analysis. The first function, io.imread, reads an image as a NumPy array; io.imshow is used to visualize an image. All other functions in this example are part of the NumPy library.

To get started, load NumPy and the input/output function of skimage (short for scikit-image) into your notebook:

```
import numpy as np
import skimage.io as io
# make Matplotlib image plotting available
# inside the notebook
%matplotlib inline
```

Before we start, a bit of background information on the science behind our image: Kacsoh et al. (2013) investigated a behavioral immune response in *Drosophila melanogaster*. Female flies lay their eggs in alcohol-laden food when confronted with parasitic wasps so that their offspring are protected from infection by the wasps. This change in oviposition behavior is mediated by neuropeptide F (NPF) and its receptor (NPFR1) in fly brains. Coupled to a reporter gene, NPF and NPFR1 can be visualized by confocal microscopy. Let's load one of the images of the study:

```
# load and show the image using the io library
# (assuming that you're working in scientific/sandbox)
image = io.imread("../data/Kacsoh2013_Drosobrain.png")
io.imshow(image)
```

If you call type(image), Python informs you that it is encoded in a NumPy array. We can therefore use NumPy's array methods to retrieve information about the image:

---

2. Note that the position (0, 0) is in the upper-left corner!

3. A black-and-white or grayscale figure has only two dimensions. Other color models, such as CMYK, would give rise to a different meaning for the third dimension.

```
image.shape
(1024, 1024, 3)
```

We see that our image measures 1024×1024 pixels and has three dimensions. The third dimension contains information on each RGB color channel. Let's look at individual channels and calculate some statistics:

```
# extract the red channel
red = image[:,:,0]
red.mean()
23.181024551391602
red.std()
30.446740821993011
red.min()
0
red.max()
255
# extract the green channel
green = image[:,:,1]
# mean value of the green channel
green.mean()
0.0
```

We see that the mean of the green channel is zero (i.e., the channel contains only zeros). You will find the same for the blue channel. This is because the confocal microscope measured only the red signal, but, nevertheless, the picture was stored as an RGB image (it could have been stored as a monochromatic image instead). RGB values range from 0 to 255. The higher the number, the brighter and more intense is the color. All three channels set to zero stands for black and all three set to 255 for white.

Let's count how many pixels are red (i.e., how many pixels display NPF and NPFR1 expression). While there are sophisticated methods to determine the background of an image, here we take the simplest approach and choose an arbitrary threshold of 100 (i.e., intensities below 100 are considered noise, all pixels with a value above this threshold are considered "red" or "expressed"). We can visually inspect how this threshold compares to the rest of the image by setting a part of our picture to 100:

```
# create a copy of the image
img_copy = image.copy()
# use Python slicing to change color values of
# part of the red channel in the image
img_copy[:, 480:500, 0] = 100
io.imshow(img_copy)
```

We now create a new array where we select only pixels that exceed the threshold, and then count their number:

```
threshold = 100
# create mask -> array of Booleans that determines whether
# pixel intensity of red channel is above threshold
mask = red > threshold
# as True == 1 and False == 0, we can plot the mask
# 0 is shown as black, 1 as white
io.imshow(mask)
# sum of mask array (i.e., number of 1s) equals
# number of pixels with NPF/NPFR1 expression
mask.sum()
```

This shows that about 37,000 pixels are above the threshold, and that the expression of NPF and NPFR1 is spatially localized.

Using the mask (i.e., setting every pixel to either 0 or 1), we lost the information on the intensity of individual pixels (i.e., strength of expression signal). However, we can show the pixels above the threshold along with their intensity by calling

```
mask2 = red * (red > threshold)
io.imshow(mask2)
mask2.sum()
```

Using NumPy arrays, we obtained a quantitative measure of NPF/NPFR1 expression from confocal images. The analysis can easily be automated and could, for example, be used to compare different treatments or genetic lineages of *D. melanogaster*.

6.2.2 Random Numbers and Distributions

To perform simulations, randomizations, or statistics, you need to access functions to draw random numbers from specific distributions, evaluate the density of a distribution at a given point, etc. This is easy to do in NumPy. For example, to sample numbers from a uniform distribution $\mathcal{U}[0,1]$, use

```
In [1]: import numpy as np
# sample two random numbers from U[0,1]
In [2]: np.random.random(2)
Out[2]: array([ 0.31622522,  0.6173434 ])
```

Similarly, you can sample integers:

```
# provide arguments maximum and how many draws
In [3]: np.random.random_integers(5, size = 10)
Out[3]: array([4, 4, 3, 5, 5, 1, 4, 3, 5, 2])
# provide arguments min, max, and how many
In [4]: np.random.random_integers(-5, -3, size = 4)
Out[4]: array([-4, -5, -5, -3])
```

or randomize the order of the elements in an array:

```
In [5]: a = np.arange(4)
In [6]: a
Out[6]: array([0, 1, 2, 3])
In [7]: np.random.shuffle(a) # shuffle in place
In [8]: a
Out[8]: array([1, 0, 2, 3])
```

You can sample random values from many common distributions, including a multivariate normal distribution:

```
# beta distribution
# parameters: alpha, beta, size
In [9]: np.random.beta(1/2, 1/2, 4)
```

```
Out[9]: array([ 0.64371741,  0.07697382,  0.04568635,
        ↳ 0.04068809])
In [10]: np.random.standard_normal(size = 3)
Out[10]: array([ 1.10453144, -0.83819954, -0.91989244])
# normal distribution
# parameters: mean, standard dev, size
In [11]: np.random.normal(10, 0.1, 4)
Out[11]: array([  9.9574825 ,  10.03459465,   9.93908575,
        ↳ 9.80264752])
# multivariate normal
# parameters vector of means, covariance matrix, size
In [12]: mus = [1, 3] # vector of means
         cov = [[1, 0.3], [0.4, 2]] # covariance matrix
         np.random.multivariate_normal(mus, cov, 3)
Out[12]: array([[-0.62639962,  1.8592831 ],
               [ 1.46134304,  4.81153371],
               [ 0.03715781,  1.41388511]])
```

An interesting fact about computers and random numbers: technically, these should be called "pseudorandom" numbers, because they are actually generated using a deterministic—yet complicated—algorithm. As we mentioned in section 4.7.3, the generation of these numbers starts with a *seed*, which you can set to reproduce exactly the (random!) outcome of a simulation. For example, if we sample two numbers from a uniform distribution,

```
In [13]: print(np.random.random(1))
         print(np.random.random(1))
[ 0.3731692]
[ 0.41573047]
```

we should always observe different values. However, you can set the seed, and then restart the sequence of random numbers from the top:

```
In [14]: np.random.seed(10)
         print(np.random.random(1))
         print(np.random.random(1))
         np.random.seed(10)
         print(np.random.random(1))
         print(np.random.random(1))
```

```
[ 0.77132064]
[ 0.02075195]
[ 0.77132064]
[ 0.02075195]
```

This is useful in at least two cases. First, suppose that you are running a set of simulations, but that every so often you incur an error. It is difficult to debug the code, as the situation triggering the error is rare. In this case, you can set the seed to an arbitrary number, run the simulation, and keep changing the number you use to seed the random number generator until you are able to reproduce the problem. Now you can fix your bug more easily, as you can reproduce the error as many times as needed.

Setting a seed is also useful when you are drawing figures. Suppose that you are showing the results of simulations, and that you want to include a figure showing the results in your paper. Then you can set the seed to an arbitrary number, so that while you are working on making the figure pretty and clear, rerunning the code will yield exactly the same results.

6.2.3 Linear Algebra

Having explored a bit of NumPy, we turn to SciPy. Because SciPy is very large, it's been divided into subpackages. Typically you will import the main package, along with one or more of the subpackages. For example, to access all functions related to linear algebra, you would type

```
In [1]: import numpy as np
        import scipy
        import scipy.linalg
```

Now you can operate on matrices, accessing functions meant to calculate eigenvalues and eigenvectors, inversion, decompositions, etc. For example,

```
# create a 3x3 matrix with random numbers
In [2]: M = scipy.random.random((3, 3))
In [3]: M
Out[3]:
array([[ 0.33037408, 0.88723624, 0.58328634],
```

```
     [ 0.52411447, 0.32525968, 0.32139096],
     [ 0.0320376 , 0.75709776, 0.47862879]])
# calculate eigenvalues
In [4]: scipy.linalg.eigvals(M)
Out[4]:
array([-0.26456383+0.j, 0.04759864+0.j, 1.35122774+0.j])
# determinant
In [5]: scipy.linalg.det(M)
Out[5]: -0.01701584715734834
# inverse
In [6]: scipy.linalg.inv(M)
Out[6]:
array([[ 5.15082966, -0.99601089, -5.60831215],
       [ 14.13739072, -8.19468158, -11.72610247],
       [-22.70738257, 13.02906339, 21.01311698]])
# product
In [7]: np.dot(M, scipy.linalg.inv(M))
Out[7]:
array([[ 1.00000000e+00, 0.00000000e+00, 1.77635684e-15],
       [ 0.00000000e+00, 1.00000000e+00, 0.00000000e+00],
       [ 1.77635684e-15, -8.88178420e-16, 1.00000000e
          ↳ +00]])
```

Note that, from a mathematical standpoint, the last operation should have returned the identity matrix (i.e., with 1s on the diagonal, and 0s everywhere else). However, we can see small numbers showing up in the off-diagonal elements. These are due to the fact that computers perform calculations with limited precision.[4]

6.2.4 Integration and Differential Equations

The subpackage integrate contains functions to perform numerical integration, and to numerically solve differential equations. For example, evaluate the left-hand side of the identity

$$\frac{22}{7} - \int_0^1 \frac{(x^2 - x)^4}{1 + x^2} \, dx = \pi \tag{6.1}$$

4. The branch of mathematics dealing with these problems is called numerical analysis. For a short, practical tutorial see computingskillsforbiologists.com/floatingpoint.

by integrating numerically.

```
In [1]: import numpy as np
        import scipy.integrate
# define the function to integrate
In [2]: def integrand(x):
            return ((x ** 2 - x) ** 4) / (1 + x ** 2)
# function, from, to
In [3]: my_integral = scipy.integrate.quad(integrand, 0, 1)
# tuple: (result, precision)
In [4]: my_integral
Out[4]: (0.0012644892673496185, 1.1126990906558069e-14)
In [5]: 22 / 7 - my_integral[0]
Out[5]: 3.141592653589793
```

Note that the function quad returns a tuple, containing the result as well as the precision obtained (you are looking for a very small number, meaning success!).

Similarly, you can use quad to numerically solve differential equations. For example, the dynamics of tumors is often modeled using the Gompertz differential equation

$$\frac{dx(t)}{dt} = \alpha x(t) \log\left(\frac{K}{x(t)}\right). \tag{6.2}$$

We can solve the equation numerically in Python, once the initial conditions $(x(0) = \frac{1}{10})$, and the parameters $(K = 5, \alpha = \frac{1}{2})$ are specified:

```
import numpy as np
import scipy
import scipy.integrate

# define function, along with the parameters
def Gompertz(x, t, alpha, K):
    dxdt = x * alpha * np.log(K / x)
    return dxdt

# initial conditions
x0 = 0.1
```

```
# parameters
alpha = 1 / 2
K = 5

# set times where we want to evaluate the solution
ts = np.arange(0, 10, 0.05)

# solve the equation
y = scipy.integrate.odeint(Gompertz, x0, ts, args =
    ↳ (alpha, K))
```

Now y contains the solution evaluated at all the times specified in ts. If you want to plot the solution, run the following code:

```
# optional: plot the solution
import matplotlib.pyplot as plt
plt.plot(ts, y[:,0])
plt.xlabel("Time")
plt.ylabel("x(t)")
plt.show()
```

Similarly, you can use the odeint function from SciPy to integrate systems of ordinary differential equations.

Intermezzo 6.1

(a) Write code to perform the integral

$$24 \int_0^1 \frac{\log^2 x}{x} \log \frac{1+x}{1-x} \, dx.$$

Is the answer π^4?

(b) The replicator equation (Taylor and Jonker, 1978; Schuster and Sigmund, 1983) is central to the study of evolutionary game theory. There are n types of individuals (e.g., alleles, species); the proportion of individuals of type i at time t is $x_i(t)$, with $\sum_1^n x_i(t) = 1$; the "fitness" of each type is given by $f_i(x)$ and the average fitness by $\phi(x) = \sum_1^n x_i f_i(x)$. With this notation, the replicator equation can be written as

$$\frac{dx_i(t)}{dt} = x_i \left(f_i(x) - \phi(x) \right). \tag{6.3}$$

Integrate the dynamics using the function `odeint` in `scipy.integrate`: start at $t = 0$, end at $t = 100$, and output the $x_i(t)$ at $t = 0$, $1, 2, \ldots, 100$. Consider the case of three types of individuals where the fitness of each type is given by the ith component of the vector $Ax(t)$, where A is a payoff matrix and $x(t)$ is the vector of proportions. Use the payoff matrix defining a rock–paper–scissors game:

$$A = \begin{bmatrix} 0 & 1 & -1 \\ -1 & 0 & 1 \\ 1 & -1 & 0 \end{bmatrix}.$$

What happens if you start the system at $x(0) = [\frac{1}{4}, \frac{1}{4}, \frac{1}{2}]^t$? What if you start at $x(0) = [\frac{1}{3}, \frac{1}{3}, \frac{1}{3}]^t$?

Supposing that you've stored the results in the two-dimensional array x, here's how you can plot the evolution of the system in time:

```
%matplotlib inline # embed plots in notebook
import matplotlib.pyplot as plt
plt.plot(t, x[:, 0], label = "x1")
plt.plot(t, x[:, 1], label = "x2")
plt.plot(t, x[:, 2], label = "x3")
plt.legend(loc = "best")
plt.xlabel("t")
plt.grid()
plt.show()
```

6.2.5 Optimization

There are many other subpackages for SciPy, and we can highlight only a few. We want to conclude this brief exposition with a subpackage that can be very helpful when one wants to find the parameter(s) for which a function is maximized (likelihoods anyone?). These methods can also be used to find the solutions of complicated equations.

For example, in epidemiology, at the end of an outbreak one wants to find the proportion of the population that did not experience a given disease. This proportion is typically called S_∞, and to calculate it one needs to solve the transcendental equation

$$S_\infty = \exp\left(-\frac{1 - S_\infty}{a/r}\right), \tag{6.4}$$

where a is the contact rate (e.g., for Ebola, ≈ 0.16) and r the recovery rate (e.g., for Ebola, $\approx \frac{1}{13}$). We can solve numerically by initially setting S_∞ to an arbitrary, small value, and trying to minimize $\exp\left(-\frac{1-S_\infty}{a/r}\right) - S_\infty$ using the root function:

```python
import numpy as np
import scipy.optimize


def my_eq(Sinf, a, r):
    return np.exp(-Sinf / (a / r)) - Sinf
```

Testing the function, we see that the value of S_∞ should be close to 0.7:

```python
In [2]: my_eq(0.6, 0.16, 1/13)
Out[2]: 0.14941562827121879
In [3]: my_eq(0.7, 0.16, 1/13)
Out[3]: 0.014238412339694917
In [4]: my_eq(0.71, 0.16, 1/13)
Out[4]: 0.00081281502599994671
```

In fact, the numerical solution is very close to 0.71: at the end of the epidemic outbreak, only about 29% of the population has experienced the disease:

```python
In [5]: scipy.optimize.root(my_eq, 0.01, args=(0.16, 1 /
    ↳ 13))
Out[5]:
        x: array([ 0.71060582])
      fun: array([  1.11022302e-16])
     fjac: array([[-1.]])
  message: 'The solution converged.'
  success: True
      qtf: array([ -4.29653313e-10])
     nfev: 7
   status: 1
        r: array([ 1.34163768])
```

where x is the array containing the result and fun is the value of the function at that point; the message states, "The solution converged." if the search was successful.

In summary, SciPy makes your work as a scientist much easier: remember to check whether a canned solution to your problem exists, before spending time and effort reinventing the wheel.

6.3 Working with pandas

pandas is the Python Data Analysis Library, introducing a data structure similar to the data.frame in R.[5] pandas provides two main data structures: Series, meant to store a one-dimensional array, and DataFrame, which contains data organized in rows and columns, as in a spreadsheet. The data stored in a Series are all of the same type; in a DataFrame, each column can be of a different type. pandas provides useful functions to manipulate data sets, calculate statistics, and plot results.

As always, we start by importing the package:

```
In [1]: import pandas
In [2]: import numpy as np # typically, both are needed
```

For our testing, we are going to import a .csv file containing a "plumage score" for male and female birds of several species. The method and the data are from Dale et al. (2015). Using pandas, we are going to import the file (start Jupyter from your scientific/sandbox directory):

```
# read CSV into pandas DataFrame
In [3]: data = pandas.read_csv("../data/Dale2015_data.csv")
```

This function creates a DataFrame object, made of rows and columns. You can specify the delimiter (e.g., sep = ";" for semicolon), change the text encoding (e.g., encoding = "latin1"), etc. The function pandas.read_excel allows you to import Excel files directly into Python. Let's explore our data set.

Use the attribute shape to determine the numbers of rows and columns of the DataFrame:

---

5. See chapters 8 and 9 for an introduction to R.

```
In [4]: data.shape
Out[4]: (5831, 7)
```

To see the first few rows in your data set, use the method head. Equivalently, tail returns the last few lines. To access the names of the columns, type:

```
In [5]: data.columns
Out[5]: Index(['Scientific_name', 'English_name', '
    ↳ TipLabel', 'Female_plumage_score', '
    ↳ Male_plumage_score'], dtype='object')
```

which returns an Index object with the column labels.

You can combine columns to create new columns:

```
# create column with sum of plumage scores
In [6]: data["Sum_scores"] = data["Female_plumage_score"]
    ↳ + data["Male_plumage_score"]
# run data.head() to see the result
```

You can also create a new column with a single operation:

```
# add a column with a constant
In [7]: data["Study"] = 1
# use NumPy function to add a column of random numbers
# shape[0] provides the number of rows
In [8]: data["Rnd"] = np.random.random(data.shape[0])
```

To remove columns from the data, use del or drop:

```
# remove a single column
In [9]: del(data["Sum_scores"])
# remove multiple columns
In [10]: data.drop(["Rnd", "Study"], axis = 1, inplace =
    ↳ True)
```

A method with option axis = 1 will act along columns, while axis = 0 acts along rows. The argument inplace = True means that the columns are removed directly and irrevocably from the data.

There are several ways of accessing data in a DataFrame: by column label,[6] row index number, or specific *x,y* locations:

```
# select data by column label
# select first three rows of output
# remember: noninclusive, 0-based indexing;
# row "3" is not included!
In [11]: data["Scientific_name"][:3]
Out[11]:
0 Abroscopus albogularis
1 Abroscopus schisticeps
2 Abroscopus superciliaris
Name: Scientific_name, dtype: object
# column names can be specified using a dot
In [12]: data.Scientific_name[:3]
Out[12]:
0 Abroscopus albogularis
1 Abroscopus schisticeps
2 Abroscopus superciliaris
Name: Scientific_name, dtype: object
```

The DataFrame methods loc and iloc select specific rows and columns without *chaining* multiple selections (e.g., data[column][row] as seen above). While loc uses row and column labels for selection, iloc expects integers that correspond to the positions of rows and columns:

```
# select rows by index label
# the row named "3" is included!
# select columns by their label
# (multiple labels within list)
In [13]: data.loc[:3, ["Scientific_name", "English_name"]]
Out[13]:
        Scientific_name English_name
0 Abroscopus albogularis Rufous-faced Warbler
```

---

6. Remember that you can hit Tab to autocomplete the column names.

```
1 Abroscopus schisticeps Black-faced Warbler
2 Abroscopus superciliaris Yellow-bellied Warbler
3 Acanthagenys rufogularis Spiny-cheeked Honeyeater
# select rows with Scientific_name Zosterops mouroniensis
In [14]: data.loc[data.Scientific_name == "Zosterops
    ↳ mouroniensis"]
Out[14]:
Scientific_name English_name TipLabel Female_plumage_score
    ↳  Male_plumage_score
5801 Zosterops mouroniensis Mount Karthala White-eye
    ↳ Zosterops_mouroniensis 47.916667 47.5
# select subset by x,y position (zero-based!)
# select third row, second column
In [15]: data.iloc[2, 1]
Out[15]: 'Yellow-bellied Warbler'
```

We can even select rows based on only part of the cell content:

```
# select the column Scientific_name of all rows that
# contain "Bowerbird" in column English_name;
# show first three rows of output
In [16]: data[data.English_name.str.contains("Bowerbird")
    ↳ ]["Scientific_name"][:3]
Out[16]:
188 Amblyornis flavifrons
189 Amblyornis inornata
190 Amblyornis macgregoriae
Name: Scientific_name, dtype: object
[...]
```

Having shown how to select particular columns, we show how to filter rows based on their content:

```
# select rows with Male_plumage_score larger than 65
In [17]: high_male_score = data[data["Male_plumage_score"]
    ↳ > 65]
```

You can also concatenate multiple conditions with Boolean operators; for example, we extract highly sexually dimorphic species by finding those in which males have a high plumage score, and females have a low score:

```
In [18]: highly_dimorphic = data[(data.Male_plumage_score
    ↳ > 70) & (data.Female_plumage_score < 40)]
```

One important feature of pandas is that many commands return a *view*, as opposed to a copy of the data. A simple example:

```
In [19]: high_male_score["Qualitative_score"] = "High"
[...] SettingWithCopyWarning:
A value is trying to be set on a copy of a slice from a
    ↳ DataFrame.
[...]
```

pandas raises a warning because high_male_score is not a new DataFrame, independent of data; rather, it is a view of a "slice" of the original data. While the behavior of this command is what you would expect (the column Qualitative_score is added to high_male_score, but not to data), this is not the correct way to proceed. When you want to take a subset of the data, and separate it from the original data, you need to copy it:

```
In [20]: high_male_score = data[data["Male_plumage_score"]
    ↳ >  65].copy()
In [21]: high_male_score["Qualitative_score"] = "High"
```

The reason behind this behavior is that pandas can be used to analyze large data sets. In most cases, you want to take a subset to perform some operations that do not alter the data—in which case copying the data by default would be costly in terms of memory. Unsurprisingly, the idea of views is taken from databases (covered in chapter 10), where it arises for the exact same reason.

Once you have selected your data, it is easy to calculate summary statistics:

```
In [22]: data.Male_plumage_score.mean()
Out[22]: 51.009189390042877
In [23]: data.Male_plumage_score.median()
```

```
Out[23]: 49.72222222
In [24]: data.Male_plumage_score.std()
Out[24]: 8.2006629346736908
```

A useful feature is that you can produce nice plots for exploratory analysis. If you have not done it already, load the Matplotlib library and type

```
In [25]: %matplotlib inline
```

to tell Jupyter to include the plots directly in the notebook. Now type

```
In [26]: data.Male_plumage_score.hist()
```

and you should see a histogram of the data! Similarly, to produce a scatter plot, use

```
In [27]: data.plot.scatter(x = "Male_plumage_score", y = "
         ↳ Female_plumage_score")
```

To draw a box plot displaying the distributions of plumage scores for males and females, type

```
In [28]: data[["Male_plumage_score", "Female_plumage_score
         ↳ "]].plot.box()
```

This introduction presents only a few functions of the quite comprehensive pandas library. If you want to perform exploratory analysis of large data sets in Python, we recommend that you master this package. Section 6.7 provides pointers to resources to further your understanding. Many of the ideas behind the pandas package are taken from R, which we will explore in chapters 8 and 9.

Intermezzo 6.2

Gächter and Schulz (2016) performed a provocative experiment to study intrinsic honesty in different countries. Groups of students were asked to perform two rolls of a fair die, and to report the result of the first roll. They were paid an amount of money proportional to the reported number, with the exception that they were given no money when they reported rolling a 6.

The subjects knew of the monetary reward, and that their rolls were private—the experimenters could not determine whether they were telling the truth or not. If everybody were to tell the truth, we would expect that each claim (from 0 to 5 monetary units) would be equally represented in the data, with a proportion of $\frac{1}{6} = 0.1\overline{6}$. Countries where cheaters were more abundant would have a higher proportion of subjects claiming a reward of 5 units and a lower proportion of those claiming 0 units.

1. Load the file (data/Gachter2016_data.csv) using pandas. Which country reported the smallest frequency of Claim == 0 (meaning fewest honest players)? Which the highest?
2. Now calculate the reported frequency of rolling the number 5 (which would lead to a claim of 5 units) for each country. Which country has the lowest frequency (most honest players)? Which the highest? Notice that the data report cumulative frequencies; to obtain the frequency of rolling a 5, you need to subtract the cumulative frequency of claiming 4 monetary units from 1.0.

6.4 Biopython

The Biopython project provides many standardized bioinformatics tools which, for example, facilitate the analysis and visualization of sequence data, the interface with data repositories, the parsing of popular file formats, and the integration of programs such as BLAST or Primer3.

Biopython is not part of the standard Python library and needs to be installed. You can find instructions in CSB/scientific/installation/install.md.

6.4.1 Retrieving Sequences from NCBI

Many of the popular biological databases offer an application programming interface (API) that allows information to be accessed programmatically. Instead of manually accessing the website, entering search terms, and clicking your way to the desired data set, you can write a script that automatically queries the database and retrieves the data. The data are downloaded in a structured format, such as Extensible Markup Language (XML), making it both human readable and machine readable. Using APIs automates your work flow, making it easy to scale up your analysis and facilitating the analysis within the Python environment.

The National Center for Biotechnology Information (NCBI) not only offers an extensive API, but also the Entrez Programming Utilities (E-utilities)—a set of server-side programs to search and retrieve information.[7] Biopython offers functions to interact directly with E-utilities. Let's see how it works by retrieving information about the inquisitive shrew mole (*Uropsilus investigator*):

```
# import the package
from Bio import Entrez
# provide your e-mail address to let NCBI know who you are
Entrez.email = "me@bigu.edu"
handle = Entrez.esearch(db = "nuccore",
        term = ("Uropsilus investigator[Organism]"))
```

The function Entrez.esearch allows us to search any of the databases hosted by NCBI,[8] returning a *handle* to the results. A handle is a standardized "wrapper" around text information. Handles are useful for reading information incrementally, which is important for large data files. Handles are often used to pass information to a parser, such as the function Entrez.read:

```
record = Entrez.read(handle)
handle.close()
# record is a dictionary, we can look at the keys
record.keys()
dict_keys(['Count', 'RetMax', 'RetStart', 'IdList', '
    ↳ TranslationSet', 'TranslationStack', '
    ↳ QueryTranslation'])
# your output may look different:
# dictionaries have no natural order
```

The Entrez.read parser breaks the retrieved XML data down into individual parts, and transforms them into Python objects that can be accessed individually. Let's see how many sequences are available in the nucleotide database for our search term, and access the record IDs:[9]

---

7. Other websites might not provide such an extensive API, but you can still extract information by using the module urllib2 and parsing structured data with the module BeautifulSoup.

8. Find a list of available databases at computingskillsforbiologists.com/entrezids.

9. Note that NCBI returns only 20 IDs by default to keep traffic on its server low. If you need all IDs, call Entrez.esearch again and set retmax to the maximum number of IDs (here 71).

```
record["Count"]
'71'
# retrieve list of GenBank identifiers
id_list = record["IdList"]
print(id_list)


['524853022', '555947199', '555947198', ... , '555946814']
```

Note that your counts and IDs might differ if more information about the inquisitive shrew mole has been uploaded since we ran our query.

Now that we know what is available (using Entrez.search) we can fetch our sequence data using Entrez.efetch. We retrieve the first 10 sequences in FASTA format and save them to a file:

```
# always tell NCBI who you are
Entrez.email = "me@bigu.edu"
handle = Entrez.efetch(db = "nuccore",
                       rettype = "fasta",
                       retmode = "text",
                       id = id_list[:10])
# set up a handle to an output file
out_handle = open("Uropsilus_seq.fasta", "w")
# write obtained sequence data to file
for line in handle:
    out_handle.write(line)
out_handle.close()
handle.close()
```

6.4.2 Input and Output of Sequence Data Using SeqIO

Next, we use the module SeqIO to manipulate our sequences and obtain more information about our *U. investigator* results:

```
from Bio import SeqIO
handle = open("Uropsilus_seq.fasta", "r")
```

```
# print ID and sequence length
for record in SeqIO.parse(handle, "fasta"):
    print(record.description)
    print(len(record))
handle.close()
KC759121.1 Uropsilus investigator isolate Deqin control
    ↳ region, partial sequence; mitochondrial
491
KF778086.1 Uropsilus investigator voucher KIZ:020539
    ↳ polycomb ring finger oncoprotein (BMI1) gene, 3' UTR
313
KF778085.1 Uropsilus investigator voucher KIZ:020527
    ↳ polycomb ring finger oncoprotein (BMI1) gene, 3' UTR
313
[...]
```

SeqIO.parse returns a SeqRecord Python object that comes with several methods. Type record. and hit the Tab key in your Jupyter notebook to obtain a list of methods. Let's select only the records of the BMI1 gene and shorten our sequences before writing to a new file:

```
import re
output_handle = open("Uropsilus_BMI1.fasta", "w")
for record in SeqIO.parse("Uropsilus_seq.fasta", "fasta"):
    # find BMI1 sequences
    if re.search("BMI1", record.description):
        print(record.id)
        # shorten sequence by Python slicing
        short_seq = record[:100]
        SeqIO.write(short_seq, output_handle, "fasta")
output_handle.close()
```

So far, we have worked exclusively with the FASTA file format. The SeqIO module can handle several other formats, and convert one into another.[10]

---

10. See the SeqIO documentation for more details: biopython.org/wiki/SeqIO.

6.4.3 Programmatic BLAST Search

The Basic Local Alignment Search Tool (BLAST) finds regions of similarity between biological sequences. Biopython provides a module to conveniently run a BLAST search against online databases:[11]

```
# NCBIWWW allows programmatic access to NCBI's BLAST
    ↳ server
from Bio.Blast import NCBIWWW
# retrieve sequences using SeqIO
handle = open("Uropsilus_BMI1.fasta", "r")
# convert SecRecord into a list for easy access
records = list(SeqIO.parse(handle, "fasta"))
# retrieve fourth sequence
print(records[3].id, " ", records[3].seq)


KF778083.1  TATTATGCTGTTTTGTGAACCTGTAGAAAACAAGTGCT[...]
```

We are ready to run our BLAST search against the NCBI nucleotide database (other options are blastp, blastx, tblastn, and tblastx):

```
# always tell NCBI who you are
Entrez.email = "me@bigu.edu"
# NCBIWWW.qplast requires three arguments:
# program, database, sequence
result_handle = NCBIWWW.qblast("blastn", "nt", records[3].
    ↳ seq)
# set up output file
save_file = open("my_blast.xml", "w")
# write results to output file
save_file.write(result_handle.read())
save_file.close()
result_handle.close()
```

---

11. You can also set up BLAST on your local computer and create your own reference database, which is useful for unpublished genome sequences, etc. See NCBI's website for setup instructions: computingskillsforbiologists.com/downloadblast.

By default, the BLAST query returns an XML file that we parse using the NCBIXML parser:

```
from Bio.Blast import NCBIXML
result_handle = open("my_blast.xml")
# use NCBIXML.read if you run BLAST for one sequence
# or NCBIXML.parse for multiple sequences
blast_records = NCBIXML.read(result_handle)
```

Next, we loop through the individual alignment hits and retrieve some basic information about results that have a good match (i.e., low E value) to our *Uropsilus* BMI1 query sequence and are more than 3000 nucleotides long:

```
E_VALUE_THRESH = 0.04
for alignment in blast_records.alignments:
    for hsp in alignment.hsps:
        if hsp.expect < E_VALUE_THRESH and alignment.
            ↳ length > 3000:
            print("****Alignment****")
            print("sequence:", alignment.title)
            print("length:", alignment.length)
            print("E value:", hsp.expect)
            print(hsp.query[0:75] + "...")
            print(hsp.match[0:75] + "...")
            print(hsp.sbjct[0:75] + "...")

****Alignment****
sequence: gi|1304911126|ref|XM_006933246.4| PREDICTED:
    ↳ Felis catus BMI1 proto-oncogene, polycomb ring
    ↳ finger (BMI1), transcript variant X3, mRNA
length: 3523
E value: 2.25861e-42
TATTATGCTGTTTTGTGAACCTGTAGAAAACAAGTGCTTTTTATC...
|||||||||||||||||||||||||||||||||||||||||||||||...
TATTATGCTGTTTTGTGAACCTGTAGAAAACAAGTGCTTTTTATC...
[...]
```

6.4.4 Querying PubMed for Scientific Literature Information

Last but not least, we query NCBI's scientific literature database, PubMed. We can, for instance, obtain information on specific journals, and search for authors or specific terms in abstracts and titles.[12] Here, we want to find the latest information on the gene *Spaetzle* in *Drosophila*. We first query the database and then extract sentences that contain the word *Spaetzle*:

```
Entrez.email = "me@bigu.edu"
handle = Entrez.esearch(db = "pubmed",
                        term = ("spaetzle[Title/Abstract]
                            ↳ AND Drosophila[ALL]"),
                        usehistory = "y")
# parse results and convert to Python dictionary
record = Entrez.read(handle)
handle.close()
# how many hits were found?
record["Count"]

'13'
# 13 records found that contained the word "spaetzle"
# and "Drosophila" (at time of preparation of this book)
```

In our `Entrez.esearch` call, we turned the `usehistory` option on. This option is recommended for complex or large queries. NCBI keeps track of our activities and we can reference a previous query up to eight hours later by providing "WebEnv" and "QueryKey" as parameters in our next E-Utilities call.

```
# store WebEnv and QueryKey in variables for later
webenv = record["WebEnv"]
query_key = record["QueryKey"]
```

Now that we know what's available, we retrieve data from the PubMed database by calling `Entrez.efetch` and write the results to a file. The query can be shortened by reusing our `WebEnv` and `WebKey` information:

---

12. Go to computingskillsforbiologists.com/pubmedtags for a list and description of all search options.

```
Entrez.email = "me@bigu.edu"
handle = Entrez.efetch(db = "pubmed",
                       rettype = "medline",
                       retmode = "text",
                       webenv = webenv,
                       query_key = query_key)
out_handle = open("Spaetzle_abstracts.txt", "w")
data = handle.read()
handle.close()
out_handle.write(data)
out_handle.close()
```

Using some regular expression magic,[13] we extract all sentences containing the word *Spaetzle*:

```
import re
with open("Spaetzle_abstracts.txt") as datafile:
    pubmed_input = datafile.read()
    # delete newlines followed by 6 white spaces
    # to have titles and abstracts on one line
    pubmed_input = re.sub(r"\n\s{6}", " ", pubmed_input)
    for line in pubmed_input.split("\n"):
        if re.match("PMID", line):
            PMID = re.search(r"\d+", line).group()
        if re.match("AB", line):
            spaetzle = re.findall(r"([^.]*?Spaetzle[^.]*\.)
                ↳ ", line)
            # don't print if list of matches is empty
            if spaetzle:
                print("PubMedID: ", PMID, " ", spaetzle)
```

You just retrieved all sentences from titles and abstracts in PubMed that contain the keywords *Spaetzle* and *Drosophila*. While you can achieve the same task relatively quickly in the graphical web interface, the programmatic approach is a lot more efficient when you want to learn about another 20 proteins and possibly repeat the search a year later. Slightly modify the script

---

13. To become a regex magician, see chapter 5.

(e.g., with a loop that cycles through your keywords of interest) and you can immediately start reading relevant information instead of spending your time clicking through web pages and manually pasting results to different files.

6.5 Other Scientific Python Modules

There are hundreds of packages and modules for writing scientific software in Python. Unfortunately, we can introduce only a handful, but want to mention at least a few others that might facilitate your research:

Matplotlib This is one of the most popular packages dedicated to plotting. It is used by pandas, and it interfaces well with NumPy and SciPy. The package seaborn[14] improves the defaults and extends the features of Matplotlib. Alternatively, ggpy[15] is a port of the popular ggplot2 package for R.[16]

rpy2 A package[17] to interface Python and R. You can execute R commands within Python, harnessing the specific advantages of each language, thereby avoiding pointless discussions on which language is superior.

SymPy A symbolic manipulation package,[18] allowing you to do math within Python. You can take derivatives, factorize polynomials, perform integrals, solve equations, etc.

OpenCV A very good library for computer vision.[19] You can use it for advanced image and video processing.

6.6 Exercises

6.6.1 Lord of the Fruit Flies

Suppose you need information on how to breed *Drosophila virilis* in your laboratory and you would like to contact an expert. Conduct a PubMed query on who has published most contributions on *D. virilis*. This person might be a good researcher to contact.

1. Identify how many papers in the PubMed database have the words *Drosophila virilis* in their title or abstract. Use the usehistory argument so you can refer to this search in the next step.

---

14. computingskillsforbiologists.com/seaborn.
15. computingskillsforbiologists.com/ggpy.
16. Covered in chapter 9.
17. computingskillsforbiologists.com/rpy2.
18. sympy.org.
19. computingskillsforbiologists.com/opencv.

2. Retrieve the PubMed entries that were identified in step (1).
3. Count the number of contributions per author.
4. Identify the five authors with the most contributions.

6.6.2 Number of Reviewers and Rejection Rate

Fox et al. (2016) studied the effects on the outcome of papers of the genders of the handling editors and reviewers. For the study, they compiled a database including all the submissions to the journal *Functional Ecology* from 2004 to 2014. Their data are reported in data/Fox2016_data.csv. Besides the effects of gender and bias in journals, the data can be used to investigate whether manuscripts having more reviewers are more likely to be rejected. Note that this hypothesis should be tested for reviewed manuscripts, that is, excluding "desk rejections" without review.

1. Import the data using pandas, and count the number of reviewers (by summing ReviewerAgreed) for each manuscript (i.e., unique MsID). The column FinalDecision contains 1 for rejection, and 0 for acceptance. Compile a table measuring the probability of rejection given the number of reviewers. Does having more reviewers increase the probability of being rejected?
2. Write a function to repeat the analysis above for each year represented in the database. For example, you should return

```
Year: 2009
Submissions: 626
Overall rejection rate: 0.827
NumRev   NumMs   rejection rate
0   306   0.977
1   2     0.5
2   228   0.68
3   86    0.698
4   4     0.75
```

6.6.3 The Evolution of Cooperation

Why are some animals (including humans) cooperating? What gives rise to complex social organizations and reciprocity? These fascinating questions can be studied using game theory, made popular in evolutionary biology by Maynard Smith (1982). One of the most well-studied problems in game theory is the "prisoner's dilemma": two prisoners are suspected of a crime and interrogated in separate rooms; each prisoner is given the possibility to either betray the other, or remain silent. If both remain silent (i.e., they *cooperate*), they each get 1 year in prison; if one remains silent (*cooperates*) and the other betrays (*defects*), the one who remained silent is sentenced to 3 years, while the other is let free; finally, if each betrays the other (*defects*), both receive a term of 2 years. Mathematically, one can show that if the game is played only once, defecting is the safest strategy (Nash equilibrium). But what if the game is played over and over? Then the mathematics becomes difficult, as the best choice depends on the choices of other players.

Axelrod's brilliant idea (Axelrod, 1980a) was to invite game theorists, sociologists, psychologists, and mathematicians to submit programs implementing different strategies for a game of iterated prisoner's dilemma. Each program would have access to the history of the moves played so far, and based on this would decide a move. Each submission then competed against itself, as well as against each other program, in a round-robin tournament.

1. Implement the following five strategies:
   (a) **always cooperate**
   (b) **always defect**
   (c) **random**: cooperate with probability $\frac{1}{2}$, and defect otherwise
   (d) **tit for tat**: cooperate on the first turn, then do whatever the other player did in the previous turn
   (e) **tit for two tat**: start by cooperating, and defect only if the other player has defected twice in a row

   Each strategy should be a function, accepting as input a list storing the previous turns of the game, and returning 1 for cooperate and 0 for defect.
2. Write a function that accepts the names of two strategies and plays them against each other in a game of iterated prisoner's dilemma for a given number of turns. Who wins between random and always_defect? And between random and tit_for_tat?
3. [**Advanced**] Implement a round-robin tournament in which each strategy is played against every other (including against itself) for 10 rounds of 1000 turns each. Who is the winner?[20]

---

20. In Axelrod's original tournament, tit_for_tat—despite being one of the simplest programs—won. This strategy was submitted by famed mathematical biologist—and University of Chicago alumnus—Anatol Rapoport. For a detailed account of the first tournament see Axelrod (1980a); a second tournament was also won by tit_for_tat (Axelrod, 1980b). For a biological interpretation, read Axelrod and Hamilton (1981).

6.7 References and Reading

SciPy

A short tutorial on NumPy:
  computingskillsforbiologists.com/numpytutorial.

pandas

The pandas website:
  pandas.pydata.org.

A ten-minute introductory video:
  computingskillsforbiologists.com/pandasvideo.

Ten nice features of pandas explained through examples:
  computingskillsforbiologists.com/pandasfeatures.

W. W. McKinney, *Python for Data Analysis*, 2nd edition, O'Reilly Media, 2017
  A book by the creator of the pandas library.

Biopython

The Biopython website:
  biopython.org.

The official Biopython tutorial and cookbook:
  computingskillsforbiologists.com/biopython.

Biopython tutorial by Peter Cock:
  computingskillsforbiologists.com/biopythonworkshop.

Other Packages

The Python website contains an extensive list of packages for science and numerical analysis:
  computingskillsforbiologists.com/numericscientific.

CHAPTER 7

• • • • • • • • • • •

Scientific Typesetting

7.1 What Is LaTeX?

LaTeX is a typesetting language that formats and arranges text, figures, and tables in a document. It is designed to deal with the complexity of scientific and technical writing. At its core, LaTeX is a bona fide programming language that allows you to define commands and use conditional branching to produce different document versions. In practice, you will often use only predefined commands to modify the layout of your document.

LaTeX is fundamentally different from a word processor (e.g., Microsoft Word) because it uses markup commands to achieve a specific layout of text. Markup languages originate in the age of typewriters: an editor would annotate (or mark up) the author's text with information for the printer—for instance to typeset a word in a bold font or underline a sentence. Markup languages (such as HTML or Markdown) are the digital analogs of this annotation process: the content is annotated with formatting commands, which are then processed to produce the formatted document.

The content and its markup annotation are stored in a text file with extension `.tex`—you can edit these files in any text editor. The interpretation of the markup commands (i.e., the actual typesetting) is done in a subsequent step (*compilation*)—typically producing a PDF file. The definitions of the markup commands are provided in separate style files that ship with your LaTeX installation or package.

7.2 Why Use LaTeX?

You know very well how to place text on a page using a word processor, so it may not be obvious why you should learn LaTeX—especially given its initially steep learning curve. A word processor is perfectly suited to simple, short documents. Many scientific texts, such as your thesis, publications, or grant proposals, are long and complex, containing many tables, figures, cross-references, and mathematical symbols.[1] While the above-mentioned separation of content and formatting might seem odd at first, it is actually very convenient for technical documents because it allows you to significantly change the layout of the document by making a few small changes to your source file. For example, you can change the layout of all figure captions at once without the need to find and alter each and every caption in your document individually.

Furthermore, LaTeX is invaluable for typesetting mathematics—pretty much all of the literature in mathematics and physics is written in LaTeX. It provides a syntax to typeset math that is so well structured that it allows visually impaired people to understand mathematical notation when it is read to them in LaTeX. Many other programs "understand" the LaTeX math notation system: for example, you can use it in your Jupyter notebooks (which we covered in the preceding chapters) and in your R Markdown files (which are introduced in the following chapters).

Another reason to transition to LaTeX is the way it manages bibliographies through its integrated bibliographic management tool, BibTeX. You can create a bibliographic database (a .bib file) and extract references to cite in your manuscripts. BibTeX automatically creates a consistently formatted bibliography based on style files that are often provided by the publishers of scientific journals. You can either manually build the bibliographic database, or conveniently download references from Google Scholar, Zotero, CiteULike, Mendeley, Scopus, Web Of Science, and many others.

Briefly, LaTeX is superior to word processors in its output quality, consistency, freedom, and engineering. Here are some more advantages of using LaTeX:

- The input is a text file (.tex). Hence, it has minimal memory requirements and is very portable. LaTeX compilers are freely available for any architecture: you obtain the same result on any computer (which is not true for most word processors).
- You can use version control for your text-based .tex and .bib files.
- LaTeX produces beautifully typeset documents. Even highly complex mathematical expressions look professional and consistent. In general, documents produced in LaTeX have a "professional look" that is produced automatically and is therefore difficult to obtain otherwise.
- LaTeX is free.

---

1. Unsurprisingly, this book is typeset in LaTeX.

- LaTeX is very stable and permanent. Originally, TeX was developed by Donald Knuth, a computer scientist at Stanford University, with a target audience of scientists. LaTeX is a simplified interface to TeX, developed by Leslie Lamport. The current version of LaTeX has basically not changed at all since 1994. Using LaTeX, you are not only independent of proprietary software, but you will never have problems with backward compatibility. LaTeX is very well designed and virtually bug-free.[2]

- Given the stability and smaller file sizes, LaTeX is the best choice for lengthy and complex documents (like your thesis). You can organize a long manuscript into separate files (i.e., chapters) that compile into one document.

- You can choose between several output formats such as .pdf, .html, .xml, and .rtf.

- Many scientific journals provide LaTeX templates, making the formatting of your manuscripts much quicker. This also facilitates the painful process of moving between two completely different formats if your paper is rejected. Most journals also provide bibliographic style files to format the bibliography generated by BibTeX.

- You can use the text editor of your choice. Most modern text editors support syntax highlighting for LaTeX. There are also a number of editors specifically for LaTeX such as Lyx (lyx.org) and TeXmacs (texmacs.org) for Ubuntu; TeXShop works wonderfully for OS X, and for Windows one can install MiKTeX.

- Integration with other free software such as R allows for a high level of automation.

- LaTeX has a long history in mathematics, computer science, and physics, but nowadays a growing number of new packages target biologists. You can format DNA or present protein alignments including structural features with a quality that is hard to achieve unless you are an expert in using image-editing software. Specific packages are available to display chemical structures, plot phylogenetic trees, etc.

- Many free books and manuals are available online. LaTeX is so popular and has been around for so many years that any problem you might have has been solved already—the solution can be found with a simple online search.

- Once you have set up a LaTeX document for a lab report or publication, you can turn it into a presentation. The required package is called Beamer. See section 7.8 to learn more.

---

2. Donald Knuth offers cash rewards to people who find bugs in TeX. Its spin-off, LaTeX, is equally stable.

Here are some disadvantages of using LaTeX and possible solutions:

- It has a steep learning curve. The separation of content and style is especially unfamiliar to nonprogrammers. However, the separation encourages a very structured writing process, and makes a lot of sense once you see how easy it is to change the style of a document.

- It is quite difficult to manage revisions with coauthors who are not familiar with LaTeX (possibly the most annoying problem with LaTeX)—but cloud-based solutions are on the rise, allowing you to leverage all the benefits of LaTeX while collaborating with someone who is not yet a convert. Some of these tools are presented in section 7.8.

- Typesetting tables is complex, but can be automated in most cases. There is, for instance, excellent support for R. You can use the `xtable` package to create files that serve as input to your `.tex` file. LaTeX itself provides packages that facilitate the input from `.csv` files.

- Floating objects (figures, tables) are set automatically, resulting in documents with a professional look. However, it can be difficult to force an object to a specific position.

- It is sometimes difficult to follow precisely the instructions of publishers if they were designed for a word processor. LaTeX, for instance, adjusts the number of lines on a page to look pretty, but sometimes you need an exact number of lines per page because of some draconian regulation mandated by publishers or funding agencies.

- Obtaining an exact word count for your compiled document (i.e., without the markup) can be tricky, but several solutions are available when you search online for "LaTeX word count."

- LaTeX is less suited to documents that are dominated by graphics or many font and color changes. These are better left to layout or graphic design tools.

7.3 Installing LaTeX

In the directory `latex/installation` we provide instructions on how to install LaTeX.

7.4 The Structure of LaTeX Documents

The structure of LaTeX documents is similar to what we have seen for other programming languages: each file starts with a preamble, defining the type of document, the packages to use, and detailing metadata, such as title and author of the manuscript. Here is an example:

```
\documentclass[12pt]{article}
\title{A simple \LaTeX{} document}
\author{Stefano Allesina, Madlen Wilmes}
```

3

As you can see, all LaTeX commands start with a backslash (\). A command may accept options (in square brackets) and arguments (surrounded by curly braces). In the example above, we chose the argument article for \documentclass, and specified the option 12pt (i.e., use 12 points as the default font size). We further defined the title and author of the document.

7.4.1 Document Classes

In the very first line of the .tex file above, we defined a document type by using the command \documentclass{[CLASS]}. We used the article document class, but other options are available (e.g., book, report, and letter). Each document class provides a comprehensive list of preset options, as well as a specific document structure.

For scientific publications, the document class article is the most suited, while for your thesis the memoir document class is very flexible. Chances are that your own institution or department offers a LaTeX template for theses, and that the scientific journal you choose provides a LaTeX template.

In addition to the general document class, you can set several global options. For example, to set the default size of the text to 10 points and US letter as the paper size, type \documentclass[10pt,letterpaper]{article}.

7.4.2 LaTeX Packages

In your preamble, you can also add specific packages. Loading additional packages allows you to alter the appearance of a document beyond the default formatting of your document class. Many packages ship with the standard LaTeX installation. Just type \usepackage{[PACKAGE]} in the preamble to load a package.

At the end of this chapter (section 7.6), we introduce a few packages that are of special interest to biologists. The following box lists some packages that are useful for enhancing the general formatting options of your document.

`\usepackage{color}`	Use colors for text in your document.
`\usepackage{amsmath,amssymb}`	Use American Mathematical Society formats and commands for typesetting mathematics.
`\usepackage{fancyhdr}`	Include "fancy" headers and footers.
`\usepackage{graphicx}`	Include figures in PDF, PS, EPS, GIF, PNG, and JPEG.
`\usepackage{listings}`	Typeset source code for various programming languages.
`\usepackage{rotating}`	Rotate tables and figures.
`\usepackage{lineno}`	Include line numbers.

A fair warning: Keep your document as simple as possible, and refrain from including too many packages. The default LaTeX installation offers beautiful typesetting for most standard documents. While it can be very rewarding and fun to explore the full power and flexibility of LaTeX, this can result in many unproductive hours spent beautifying your documents without adding a single line of content.

7.4.3 The Main Body

Once you have selected packages, you can start your main document with `\begin{document}`, and end it with `\end{document}`. The actual content of your document goes in between these two commands.

Let's create our first `.tex` file. You can use a dedicated LaTeX editor, or any other text editor.[3] Type the following LaTeX code and save it as `My_document.tex` in the directory `latex/sandbox`:

```
\documentclass[12pt]{article}
\title{A simple \LaTeX{} document}
\author{Stefano Allesina, Madlen Wilmes}
\date{}
```

Line 3:

---

3. Many programming editors have LaTeX support, so you can use the same editor for all your programming needs, whether you write code in Python, R, or LaTeX. Examples include gedit, emacs, Atom, Sublime, and vim.

```
6   \begin{document}
    \maketitle

9   \begin{abstract}
        Every scientific article needs a concise, clearly
        written abstract that summarizes the findings.
12      It should leave the reader with no choice but to read
        the entire paper and cite it as soon as possible.
    \end{abstract}

15
    \end{document}
```

The first command within our document environment is \maketitle. It tells the LaTeX compiler to use the information from the preamble to generate a title. The exact layout of the title depends on the document class you choose. You can suppress the date by including \date{} in your preamble.

Now you can create a PDF of your article. If you use a LaTeX editor, you can usually compile your document with the click of a button. LaTeX, however, does not rely on such an editor and you can compile directly in the terminal. Here, we follow this latter approach, as it is consistent across all platforms. In your terminal, change into the directory where your .tex file is located and type

```
$ pdflatex My_document.tex
$ pdflatex My_document.tex
```

You might wonder why you have to execute the command pdflatex twice in order to correctly compile your document. In the first run, the LaTeX engine needs to see what is in your file and take note of cross-references, citations, etc. This information is stored in an "auxiliary" file (i.e., My_document.aux), which is created in your directory. The second time around, this information is processed and integrated into your final output. In our simple example, we do not yet have cross-references, a bibliography, or an index, but you should make it a habit to compile twice.

Let's have a look at our My_document.pdf, which you can open using your system's default PDF viewer.[4]

---

4. Note that we show only part of the compiled PDF.

A Simple LaTeX Document

Stefano Allesina, Madlen Wilmes

Abstract

Every scientific article needs a concise, clearly written abstract that summarizes the findings. It should leave the reader with no choice but to read the entire paper and cite it as soon as possible.

7.4.4 Document Sections

Document sections allow you to structure a document. Different document classes provide slightly different document sections. For instance, the document class book offers a \chapter command, while the article document class offers an abstract environment.

Most structural elements are available for all document classes:

```
1  \section{A section}
   \subsection{A subsection}
   \subsubsection{A subsubsection}
4  \paragraph{A paragraph has no numbering}
   \subparagraph{A subparagraph has no numbering}
```

Additionally, the document classes book and report use

```
1  \chapter{My Chapter Heading}
```

Using an asterisk within a \section* call (or other structural element) will leave this section without numbering. It will also not be considered for the table of contents (which can be generated by including the command \tableofcontents within the document environment).

7.5 Typesetting Text with LaTeX

7.5.1 Spaces, New Lines, and Special Characters

White space in your LaTeX document is not equal to the white space in your compiled typeset document: several spaces in your text editor are treated as one space in the typeset document; several empty lines are treated as one empty line. One empty line in your TeX file defines a new paragraph. You can add extra white space in your compiled document by using the commands `\vspace{0.5in}` (vertical space) or `\hspace{20mm}` (horizontal space) for example.

Some characters have special meaning in LaTeX, and you need to *escape* them if you want to type them in your text. In order to type one of # $ % fi & _ { } ~ \, you need to add a backslash in front of the symbol, so that typing `\$` in your LaTeX document produces $ in your compiled document. The exception to this rule is the backslash (\) itself. Typing \\ in your LaTeX file produces a line break. If you want to show a \ in your compiled document, you need to use the command `\textbackslash`.

7.5.2 Commands and Environments

We have already seen several examples of LaTeX commands that start with a \ and may accept input within curly braces (e.g., `\section{}`).

Add the following LaTeX source code to your `My_Document.tex`. The code demonstrates some of the LaTeX commands that format text:

```
  \section{Introduction}
2 Every scientific publication needs an introduction.

  \section{Materials \& Methods}
5 The \textbf{Hardy--Weinberg equilibrium model}\footnote{
      Godfrey Hardy, an English mathematician, and Wilhelm
      Weinberg, a German physician, developed the concept
      independently.} constitutes the \textit{null model} of
      population genetics. It characterizes the distributions
      of \texttt{genotype frequencies} in populations that
      are \underline{not evolving}.
```

Besides commands, LaTeX also offers *environments*. These start with `\begin{[ENVIRONMENT]}` and are ended by `\end{[ENVIRONMENT]}`. Any text within the environment is subject to special formatting specified by the environment. For example, we have already used the abstract environment in our document, producing a block of centered text. The default LaTeX installation ships with a large selection of commands and environments, and many more can be loaded by the packages. Notably, you can define your own commands and environments—though we are not going to cover this feature in this introductory chapter.

7.5.3 Typesetting Math

LaTeX excels at typesetting math. No other program produces such consistent and beautiful mathematical symbols and equations. There are two options to display mathematical content: using inline mathematics (i.e., within the text), or using stand-alone, numbered equations and formulae.

Both inline and stand-alone mathematics can be introduced in multiple ways, either by using an environment call or by using a shorthand as summarized in the next box.

There are several options for typesetting inline or stand-alone math:

Inline

```
\begin{math} ... \end{math}
\( ... \)
$ ... $
```

Stand-alone:
numbered `\begin{equation} ... \end{equation}`
unnumbered `\[ ... \]`

Let's add some mathematics to our document:

```
1   We assume that $p$ is the frequency of the dominant allele
        ($A$) and  $q$ is the frequency of the recessive allele
        ($a$). Given that the sum of the frequencies of both
        alleles in a population in genetic equilibrium is 1, we
        can write
    \[ p + q = 1, \]
    hence
4   \[ (p + q)^2 = 1, \]
    which resolves to
```

```
\begin{equation}
p^2 + 2pq + q^2 = 1.
\end{equation}
In this equation, $p^2$ is the predicted frequency of
    homozygous dominant ($AA$) individuals in a population,
      $2pq$ is the predicted frequency of heterozygous ($Aa$
      ) individuals, and $q^2$ is the predicted frequency of
    homozygous recessive ($aa$) ones.
```

Compile `My_document.tex` and have a look at the beautifully typeset math!

2   Materials & Methods

The **Hardy–Weinberg equilibrium model**[1] constitutes the *null model* of population genetics. It characterizes the distributions of `genotype frequencies` in populations that are not evolving.

We assume that p is the frequency of the dominant allele (A) and q is the frequency of the recessive allele (a). Given that the sum of the frequencies of both alleles in a population in genetic equilibrium is 1, we can write

$$p + q = 1,$$

hence

$$(p + q)^2 = 1,$$

which resolves to

$$p^2 + 2pq + q^2 = 1. \tag{1}$$

[1]Godfrey Hardy, an English mathematician, and Wilhelm Weinberg, a German physician, developed the concept independently.

1

LaTeX has a full set of mathematical symbols and operators. To give you an idea, we provide the LaTeX source code on the left and the compiled result on the right:

```
Limits and summations,
subscripts and
superscripts can be
grouped using curly
brackets:
\begin{equation}
\lim_{x \to \infty}, \sum_
{i=1}^{\infty}, x^{y^{z^{2
k}}}, x_{y_{z^2}}.
\end{equation}
```

Limits and summations, subscripts and superscripts can be grouped using curly brackets:

$$\lim_{x \to \infty}, \sum_{i=1}^{\infty}, x^{y^{z^{2k}}}, x_{y_{z^2}}. \tag{7.1}$$

Of course, you can typeset Greek letters, fractions, square roots, arrows, summations, integrals, special functions, operators, and probably any other symbol that you can think of.[5]

```
\[ \Theta = \frac{\
Sigma\beta^2}{\sqrt
[3]{x} / (y \frac{k}{\
ln x})} \]

\[ \overline{abcd}\
underline{xy} \]

\[ \hat{a}\,\
overrightarrow{ky} \]

\[ \downarrow \uparrow
 \rightarrow \
Leftarrow \]

\[ \sum, \prod, \int,
\iiiint, \bigcup \]

\[ \cos, \exp, \min, \
log, \tanh \]

\[ \times, \pm, \neq,
\leq, \supset, \in, \
propto \]
```

$$\Theta = \frac{\Sigma\beta^2}{\sqrt[3]{x}/(y\frac{k}{\ln x})}$$

$$\overline{abcd}\underline{xy}$$

$$\hat{a}\,\overrightarrow{ky}$$

$$\downarrow\uparrow\rightarrow\Leftarrow$$

$$\sum, \prod, \int, \iiiint, \bigcup$$

$$\cos, \exp, \min, \log, \tanh$$

$$\times, \pm, \neq, \leq, \supset, \in, \propto$$

7.5.4 Comments

Anything following a percentage sign (%) is considered a comment and is disregarded in the compiling process. The end of the line signals the end of the comment. Remember to escape the percentage sign if you actually want to include it in your typeset text, for example,

---

5. A comprehensive list of LaTeX symbols is available in PDF format at computing skillsforbiologists.com/latexsymbols. Don't know the name of a symbol that you would like to include? Check out the Detexify application by Daniel Kirsch: computingskillsfor biologists.com/detexify.

```
% Include confidence
% intervals!
Body mass was reduced
by 9\%.
```

Body mass was reduced by 9%.

7.5.5 Justification and Alignment

The next box provides some helpful commands and environments to format text.

\\ Break a line without starting a new paragraph.

\newpage Start a new page.

\clearpage Clear the remainder of the page. Helpful after floating objects (e.g., figures and tables).

\cleardoublepage Ensure the new page starts on an odd-numbered page (e.g., chapters).

\begin{flushleft} Align the text on the left. End by \end{flushleft}.

\begin{flushright} Produce right-justified text. End by \end{flushright}.

\begin{center} Produce centered text, figures, or tables. End by \end{center}.

7.5.6 Long Documents

If you are writing a long document, such as your thesis, it makes sense to split it into semi-independent files that are easier to handle. Build a master file containing the document type, the basic settings, your personal commands and environment definitions, etc., and then create a LaTeX file for each chapter. The contents of these chapter files are the text as if you were to type it directly into the master file (i.e., chapter files have no preamble). Therefore, they cannot be compiled independently, but rather require compilation through the master file.

Finally, include (or exclude) each chapter file using the \input command in your master file. We will not set up a complex document here but reading through the following example, will help clarify the principle:

```
\documentclass{report}

% load desired packages
\usepackage{amsmath}
```

```
      \usepackage{graphicx}
6     % Specify the directory where pictures are stored
      \graphicspath{{Pictures/}}

9     \begin{document}

      \title{\textbf{Your short or long thesis title}}
12    \author{Your Name}

      % let LaTeX produce the title and table of contents
15    \maketitle
      \tableofcontents

18    \input{introduction}
      \input{methods}
      \input{results}
21    \input{discussion}

      \end{document}
```

Now suppose you need to change the order of your chapters: that's when it pays to have used LaTeX! Just change the order of the input files and recompile the master .tex file: the numbering of all the headers, figures, citations, tables, etc. are automatically adjusted.

7.5.7 Typesetting Tables

The basic environment to typeset tables is called tabular.

```
\begin{tabular}{l|r|c}
A & B & C \\
\hline
AA & BB & CC \\
\hline
AAA & BBB & CCC\\
\end{tabular}
```

A	B	C
AA	BB	CC
AAA	BBB	CCC

From the example above, you can see that cells are separated by an ampersand (&) and that a double backslash (\\) signals the end of the line. The alignment within cells is specified immediately after the \begin{tabular} command using a single letter code: l stands for left, r for right, and c for center alignment. The number of letters needs to match the number of columns in the table. A vertical bar (|) between the alignment specifications produces a vertical line to separate columns. Horizontal lines are placed with the command \hline.

Here is how to typeset tables that contain multiple-column cells and multiple-line cells:

```
multiple-column cells:

\begin{tabular}{|c|c|}
\hline
  A & B \\
\hline
  \multicolumn{2}{|c|}{CD
  }\\
\hline
\end{tabular}
```

multiple-column cells:

A	B
CD	

```
multiple-line cells:

\begin{tabular}{|c|c|p{3cm
}|}
\hline
A  & B & We fix the cell
width at 3 cm so the long
line breaks.\\
\hline
\end{tabular}
```

multiple-line cells:

A	B	We fix the cell width at 3 cm so the long line breaks.

Admittedly, typesetting tables in LaTeX is a bit strange, but remember that LaTeX is entirely text based, so all elements of a table are individually set by text commands. The good news is that you will rarely have to set large tables by hand. You can achieve a good deal of automation by running your data analysis in R and then use the packages xtable or sweave to directly output a LaTeX table that you \include in your .tex file. In this way, you do not need to manually rewrite your table just because you reanalyzed your data with slightly different parameters. If need be, there are also ways to convert a .csv to .tex.[6]

In papers and other documents, you typically include "floating" tables. Floating means that the object cannot be broken across pages, and thus LaTeX tries to place it where it looks "pretty." Each table starts with the command \begin{table} followed by a specifier that determines its position. You can use the following options:

Specifier	Where to place the floating table
h	Here: try to place the table where specified
t	At the top of a page
b	At the bottom of a page
p	On a separate page
!	Try to force the positioning (e.g., !h)

Let's add a table to our example file My_Document.tex. We include a caption by specifying the \caption command. You can modify the position of the caption by placing the command before or after the floating element:

```
\section{Results}

\subsection{Genotype Frequencies}

We estimated genotype frequencies of 1612 individuals.

\begin{table}[h]
  \begin{center}
    \caption{Frequency distribution of observed phenotypes
      .}
    \vspace{.1in}
    \begin{tabular}{lc} \textbf{Phenotype} & \textbf{
      Frequency} \\
      \hline
      $AA$ & 1469 \\
      $Aa$ & 138 \\
      $aa$ & 5 \\
      \hline
```

---

6. See tablesgenerator.com/latex_tables or the LaTeX package csvsimple.

```
17        \textbf{Total} & \textbf{1612}\\
        \end{tabular}
      \end{center}
20  \end{table}
```

Compile the .tex file and have a look at our typeset table. It should look similar to the figure below, showing part of page 2 of My_Document.pdf:

3 Results

3.1 Genotype Frequencies

We estimated genotype frequencies of 1612 individuals.

Table 1: Frequency distribution of observed phenotypes.

Phenotype	Frequency
AA	1469
Aa	138
aa	5
Total	1612

Note that LaTeX automatically numbers the table. The document class of your document will influence how tables (and other floating objects, such as figures) are numbered. While the document class book causes tables to be numbered including a chapter number (e.g., table 3.1), the document class article will lead to continuous numbering throughout the document (e.g., figure 1, figure 2, etc.).[7]

If you need a really long table (spanning multiple pages), there are specific packages for that, such as longtable. Consult its manual for detailed instructions on how to use the package.

7.5.8 Typesetting Matrices

Matrices are defined by the array environment, which is similar to the tabular environment introduced above:

---

7. If you want to change the default behavior of your document class, have a look at the chngcntr package.

```
\begin{equation}
  A = \left[
    \begin{array}{ccc}
      \alpha & \beta & \
      gamma\\
      \mathfrak a & \
      mathfrak b & \
      mathfrak c
    \end{array}
    \right]
\end{equation}
```

$$A = \left[\begin{array}{ccc} \alpha & \beta & \gamma \\ \mathfrak{a} & \mathfrak{b} & \mathfrak{c} \end{array} \right] \tag{7.2}$$

```
\begin{equation*}
  B = \left|
  \begin{array}{cc}
    a^2 & bac \\
    \frac{g}{f} & fa^3\
    sqrt{b}
  \end{array}
  \right|
\end{equation*}
```

$$B = \left| \begin{array}{cc} a^2 & bac \\ \frac{g}{f} & fa^3\sqrt{b} \end{array} \right|$$

```
\begin{equation}
  C = \left.
  \left(
    \begin{array}{ccc}
      \alpha & \beta & \
      gamma\\
      a & b & c\\
    \end{array}
    \right)
  \right|_{a = a_0}
\end{equation}
```

$$C = \left. \left(\begin{array}{ccc} \alpha & \beta & \gamma \\ a & b & c \end{array} \right) \right|_{a=a_0} \tag{7.3}$$

7.5.9 Figures

Figures are also floating objects. They can be included using the graphicx package. Depending on your installation, you might be able to include only some file formats. Typically, you should be able to use .pdf, .ps, .eps, .jpg, and .png files, among others.

In order to add a figure to My_Document.tex, we include \usepackage {graphicx} in the document's preamble and add the following LaTeX source code:[8]

```
   \subsection{Gel Electrophoresis}
2
   We identified differences in allelic composition by gel
      electrophoresis.

5  \begin{figure}[h]
     \begin{center}
       \includegraphics{electrophoresis.png}
8    \end{center}
   \caption{Gel electrophoresis of four individuals.}
   \end{figure}
```

As for tables, you can add captions that are automatically numbered. The caption of a figure can be placed above or below the figure, depending on the position of \caption with respect to \includegraphics. Let's compile and take a look at the result:

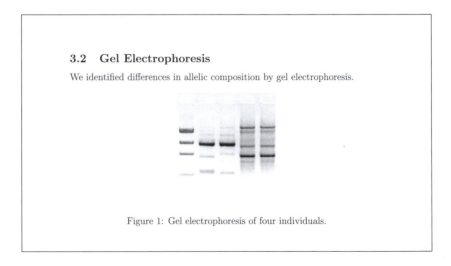

3.2   Gel Electrophoresis

We identified differences in allelic composition by gel electrophoresis.

Figure 1: Gel electrophoresis of four individuals.

---

8. To successfully compile this code, place the image electrophoresis.png (located in CSB/latex/data) in the same directory as My_Document.tex.

Browse the documentation of the graphicx package and you will see that you can modify the appearance of the figure in various ways without altering the image itself. For instance, you can specify the width, height, scale, set it at an angle, or even trim and clip margins:

```
\subsection{Gel Electrophoresis}

We identified differences in allelic composition by gel
    electrophoresis.

\begin{figure}[h]
  \begin{center}
    \includegraphics[scale=0.8, trim={0cm 0cm 2.6cm 0cm},
      clip]{electrophoresis.png}
  \end{center}
  \caption{Gel electrophoresis of one individual. A
    molecular-weight size marker is shown to the left for
    comparison.}
\end{figure}
```

The code above yields the following result:

3.2   Gel Electrophoresis

We identified differences in allelic composition by gel electrophoresis.

Figure 1: Gel electrophoresis of one individual. A molecular-weight size marker is shown to the left for comparison.

7.5.10 Labels and Cross-References

One of the strengths of LaTeX is that it will automatically label sections, equations, and floating objects. No matter how you shift everything around in your writing process, the objects will be labeled in the correct order. You can further exploit this feature by associating a label with any numbered object and cross-reference it in your text. The label is just a "tag"—it will not be printed in the document. Instead, whenever you call the command \ref{} with that label as parameter, LaTeX will provide the correct number for the object. For instance, if you wrote about figure 1 in a given sentence but you then moved it to be figure 5, LaTeX will not only automatically update the figure number, but also update all the mentions of the figure in the text.

For example, let's add a label to our gel electrophoresis subsection and to the figure that we cross-reference in the text:

```
   \subsection{Gel Electrophoresis}\label{sec:electrophoresis}
2
   We identified differences in allelic composition by gel
   electrophoresis (figure~\ref{fig:electrophoresis}).
5
   \begin{figure}[h]
     \begin{center}
8      \includegraphics[scale=0.8, trim={0cm 0cm 2.6cm 0cm},
           clip]{electrophoresis.png}
     \end{center}
     \caption{Gel electrophoresis of one individual. A
         molecular-weight size marker is shown to the left for
         comparison.}\label{fig:electrophoresis}
11 \end{figure}

   \section{Discussion}
14
   Our experiment demonstrates differences in allelic
       composition
   (section~\ref{sec:electrophoresis}).
```

7.5.11 Itemized and Numbered Lists

There are several LaTeX environments used to produce lists and descriptions:

```
\begin{itemize}
\item First note
\item Second note
\item[$\star$] Third note
\item[(a)] Fourth note
\end{itemize}
```

- First note

- Second note

* Third note

(a) Fourth note

```
\begin{enumerate}
\item First.
\item Second.
\item Third.
\end{enumerate}
```

1. First.
2. Second.
3. Third.

```
\begin{description}
\item[First] A short
description of First
\item[Second] A short
description of Second
\item[Third] A much longer
 description: no worries,
  \LaTeX{} arranges
  everything
\end{description}
```

First   A short description of First

Second A short description of Second

Third   A much longer description: no worries, LaTeX arranges everything

7.5.12 Font Styles

When we typed the Materials & Methods section of our example document (section 7.5.2), we changed the typeface in order to highlight text. We can also change its size. Here is an overview of the required commands:

```
\begin{itemize}
\item \texttt{Typewriter-
like}
\item \textbf{Boldface}
\item \textit{Italics}
\item \textsc{SmallCaps}
\item {\tiny Tiniest font}
\item {\footnotesize
Footnote-text size}
\item {\small A smaller
font than normal}
\item {\normalsize Normal
size}
\item {\large A larger
font}
\item {\Large Even larger}
\end{itemize}
```

- Typewriter-like
- **Boldface**
- *Italics*
- SMALLCAPS
- Tiniest font
- Footnote-text size
- A smaller font than normal
- Normal size
- A larger font
- Even larger

Note that the size of the text (i.e., what is considered the normalsize) is governed by the initial choice in documentclass. The other sizes are adjusted proportionally.

7.5.13 Bibliography

Managing references is essential for scientific writing. BibTeX is a citation management tool that is tightly integrated into the LaTeX typesetting system, tremendously facilitating consistent formatting of your reference list and the management of reference styles (e.g., from author–year to numbered citations).

BibTeX stores references in a text-based database, called the .bib file. It contains each reference in a specific BibTeX format. BibTeX records can be downloaded from Web of Science, Scopus, etc., or you can use software such as BibDesk, Mendeley, Papers, or EndNote to organize your bibliography and generate a .bib file. Each reference entry in the .bib file contains a specific identifier (citation key) that is used to integrate the reference into your scientific text.

Let's add references to our methods section of My_Document.tex. The file latex/data/My_Document.bib contains two entries with the keys Weinberg1908 and Hardy1908, respectively. We can now add the citations to our document using the command \cite:

```
\section{Materials \& Methods}
The \textbf{Hardy--Weinberg equilibrium model}\footnote{
    Godfrey Hardy, an English mathematician, and Wilhelm
    Weinberg, a German physician, developed the concept
    independently.} constitutes the \textit{null model} of
    population genetics. It characterizes the distributions
    of \texttt{genotype frequencies} in populations that
    are \underline{not evolving} \cite{Hardy1908,
    Weinberg1908}.
```

After inserting the appropriate citations, we also have to generate the list of references. We want to generate it at the very end of the document so we place the appropriate commands just before the \end{document} call:

```
1   \bibliography{My_Document}
    \bibliographystyle{plain}
```

The command \bibliography indicates the name and location of the corresponding .bib file. Make sure that the .bib and .tex files are in the same directory or adjust the path to the file. Note that the file name of the bibliography file is referenced without its .bib file extension. Next we include the command \bibliographystyle to control the formatting of the reference list. We use the plain style here but many journal-specific or general styles are available.

In order to compile the file, type the following into your terminal (or press the corresponding buttons in your GUI):

```
$ pdflatex My_Document.tex
$ pdflatex My_Document.tex
$ bibtex My_Document
$ pdflatex My_Document.tex
$ pdflatex My_Document.tex
```

The multiple runs are required to connect the .bib database entries with the citation commands, and then compose the reference list. Here are the parts of our compiled PDF that show the numbered citations in the text and a formatted bibliography at the end of the document:

2 Materials & Methods

The **Hardy–Weinberg equilibrium model**[1] constitutes the *null model* of population genetics. It characterizes the distributions of `genotype frequencies` in populations that are not evolving [1, 2].

We assume that p is the frequency of the dominant allele (A) and q is the frequency of the recessive allele (a). Given that the sum of the frequencies

4 Discussion

Our experiment demonstrates differences in allelic composition (section 3.2).

References

[1] Godfrey Hardy. Mendelian proportions in a mixed population. *Science*, 28:49–50, 1908.

[2] Wilhelm Weinberg. *Über den Nachweis der Vererbung beim Menschen*, volume 64. Jahreshefte des Vereins Varterländische Naturkunde in Württemberg, 1908.

For more flexibility in formatting your references, see the package `natbib`. It is not actively developed anymore but is reliable and useful, especially for your thesis—its many commands offer great flexibility on how to include citations. Many scientific journals, however, accept references only in BibTeX format: check whether the journal provides a template beforehand.

The package `biblatex` is meant to replace BibTeX and is currently gaining traction. It allows insertion of multiple bibliographies and easy modification of citations styles. Unfortunately, it is not yet widespread among scientific publishers as it is not compatible with BibTeX style files, nor does it provide a `.bbl` file, which some publishers require.

7.6 LaTeX Packages for Biologists

LaTeX's main purpose is typesetting documents, but some packages go beyond placing text on the page and make use of its programming capabilities. The following packages allow you to format sequence alignments, and plot phylogenetic trees or chemical structures. While similar results may be obtained with a graphic editor, LaTeX allows automation, better documentation of how graphics are produced, and therefore reproducibility and consistency. Any additional time to set up a figure using LaTeX is easily compensated for when the figure has to be redrawn due to changes in the data input.

7.6.1 Sequence Alignments with LaTeX

The TeXshade package is a comprehensive tool to display, shade, and label nucleotide and protein alignments (Beitz, 2000). The data input is a text file in MSF, ALN, or FASTA format—which are standard outputs of sequence alignment programs. TeXshade offers extensive options to format alignments to specific needs (e.g., to display DNA similarity or diversity, consensus sequences, structural elements in proteins, or emphasis of specific residues). Here we show a simple example. Please refer to the manual for more details on usage and commands.

A FASTA file with five amino-acid sequences, and the following seven lines of LaTeX code, produce the complex figure shown below. The sequences are taken from Malhotra et al. (2013) and provided in CSB/latex/data.[9]

```
1   \begin{texshade}{PLA2.fasta}
    \setends{1}{34..56}
    \showsequencelogo{top}
4   \feature{top}{1}{39..54}{helix}{Helix II}
    \showruler{bottom}{1}
    \hidenumbering
7   \end{texshade}
```

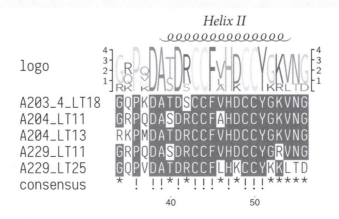

---

9. Remember to load the package in your preamble using \usepackage{texshade}.

7.6.2 Creating Chemical Structures with LaTeX

The Comprehensive TeX Archive Network (ctan.org) lists several packages for depicting chemical structures. Here is a simple example showing how to draw chemical structures using the chemfig package:

```
\setdoublesep{0.3em}
\setatomsep{1.8em}
\chemname{
\chemfig{*6((=O)-N(-CH_3)
-*5(-N=-N(-CH_3)-=)--(=O)-
N(-H_3C)-)}
}{Caffeine fuels biology.}
```

Caffeine fuels biology.

7.7 Exercises

7.7.1 Typesetting Your Curriculum Vitae

Updating your curriculum vitae (CV) every time you publish a new paper, give a talk, or receive a grant or fellowship should become a good habit. Throughout your career, you will be asked to produce your CV (in many different formats) for job and grant applications, fellowships and prizes, or for promotion and tenure. Keeping it up to date will save you a lot of time.

LaTeX is ideally suited for this task, especially because you can import all your publications from a .bib database—no need to type them up, simply download them from your favorite website.

To illustrate how easy it is to produce a good-looking, easy-to-update CV in LaTeX, in the directory latex/data/cv you will find three examples of CVs (one bare bones, one more colorful, and a really fancy one—you need to install software to be able to compile the latter one). Choose the one you like the best (or download templates from LaTeX-based websites[10]), and lay the foundations for your own CV.

---

10. For example, see the websites computingskillsforbiologists.com/resumetemplates and computingskillsforbiologists.com/latexresume.

7.8 References and Reading

Books

Frank Mittelbach et al., *The LaTeX Companion*, Addison-Wesley Professional, 2004.
Somewhat outdated but still relevant as the ultimate LaTeX reference book.

Online Resources

The Visual LaTeX FAQ: sometimes it is difficult to describe in words what you need!
computingskillsforbiologists.com/visuallatex.

Drawing formulas by hand and converting them into LaTeX:
computingskillsforbiologists.com/drawmath.

Online equation editors:
computingskillsforbiologists.com/equationeditor,
hostmath.com.

Dissertation templates:
computingskillsforbiologists.com/thesistemplates.

Cloud-based collaborative writing using LaTeX:
overleaf.com,
sharelatex.com,
authorea.com.

Manage revisions of LaTeX documents:
computingskillsforbiologists.com/latexdiff.

Beautiful presentations in LaTeX using Beamer:
computingskillsforbiologists.com/beamer.

Many scientists and designers produced beautiful templates for Beamer:
computingskillsforbiologists.com/presento,
computingskillsforbiologists.com/mtheme.

Bibliographies for biological journals:
computingskillsforbiologists.com/bibtexstyles.

Tutorials and Essays

A (Not So) Short Introduction to LaTeX 2ε:
computingskillsforbiologists.com/latexintro.

The Beauty of LaTeX:
computingskillsforbiologists.com/beautyoflatex.

Word vs. LaTeX goes "scientific":
Knauff and Nejasmic (2014) concluded that preparing simpler documents with MS Word is faster and more efficient than doing it in LaTeX. It would be interesting to see the same experiments for more complex documents.

CHAPTER 8

• • • • • • • • • • • •

Statistical Computing

8.1 Why Statistical Computing?

In chapters 3–6 we worked with Python, which is a general-purpose programming language. Many languages, on the other hand, are domain specific, and adding software to your toolbox that targets, for example, statistical/mathematical computing, will make you a more productive scientist. You can use this software to conduct statistical analysis of your data, produce graphs and figures, and manipulate large data sets.

In this chapter, we provide an overview of R. Alternatively, you might want to explore the use of MATLAB[1] (proprietary, very popular among engineers), Julia[2] (free software, very fast thanks to a just-in-time compiler), or Octave[3] (free clone of MATLAB), or even use Python in conjunction with the packages SciPy, pandas, and Matplotlib or ggplot. If you need to do much mathematical work (e.g., symbolic manipulation of variables or analytic integration and differentiation), take a look at Mathematica[4] (proprietary), or Sage[5] (free, combining many features of R and Python).

8.2 What Is R?

R is software for statistical analysis. It comes with many built-in functions and excellent graphical capabilities. The main strength of R is that it is fully programmable: you can write code in R and have the software

---

1. mathworks.com.
2. julialang.org.
3. gnu.org/software/octave.
4. wolfram.com/mathematica.
5. sagemath.org.

execute it. This means that it is very easy to automate your statistical and data analysis.

The fact that R is easy to program led to the development of thousands of packages: you can find a ready-made package for almost any statistical analysis you might want to perform, no matter how specialized your research interests might be. Because of this strength, R has become the most popular statistical software among biologists.

The main hurdle new users face when learning R is that it is based on a command-line interface: to make things happen, you write text commands in a "console," and then the program executes them. This might seem unusual if you are used to software based on a graphical user interface, where you tend to work by clicking on windows and buttons. However, the command line is what makes it easy to automate your analysis—all you have to do is collect all the commands in a text file, and then run them in R.

8.3 Installing R and RStudio

For this introduction, we are going to use RStudio, a graphical interface that simplifies the use of R by giving you immediate access to the code, the console, and the graphics.

To install R and RStudio, follow the instructions contained in `CSB/r/installation`.

You can launch RStudio by clicking on its icon, or by opening a terminal and calling `rstudio`.

8.4 Why Use R and RStudio?

The main advantage of R is that you can write scripts for all your work, instead of manually typing commands and clicking buttons. This makes your research easy to reproduce, well documented, and and easy to automate. We will learn more about the advantages of working with scripts in section 8.10.1.

Furthermore, the details of your analysis are entirely in your hands—you do not need to adapt your analysis to the available software, but rather adapt the software to your analysis.

Another advantage is that R is free software: it is free to use, but it also gives you the freedom to see the code (open source), modify it, and extend it. This strength has led to the development of hundreds of highly specialized packages for biological research.

RStudio is an integrated development environment (IDE) for R that is available for all platforms. It includes a console, and panels showing your plots, the command history, and information on your workspace. You can use the integrated editor to write your scripts. The editor highlights your R syntax. To run a part of the code, select it and press Ctrl+Enter; to run the whole script, use Ctrl+Shift+S.

8.5 Finding Help

Each command in R comes with a manual page. To access it, type ?[NAME OFCOMMAND] in the console (e.g., ?lm).

8.6 Getting Started with R

We will first explore simple operations, assignments, and data types to get familiar with the R console. A "greater than sign" (>) at the beginning of the line in the console means that R is ready to accept your input. You can navigate the history of previously typed commands by using the arrows on your keyboard. A comment starts with the hash mark (#), and does not require a closing symbol.

We can use R as an oversized calculator:

```
> 1.7 * 2
[1] 3.4
> 12 / 5
[1] 2.4
> 2.1 ^ 5
[1] 40.84101
> log(10)
[1] 2.302585
> log10(10)
[1] 1
> sqrt(9)
[1] 3
> q() # quit the R session
```

The following box provides a list of arithmetic and logical operators in R:

Operator	Description
+	Addition
–	Subtraction
*	Multiplication
/	Division
^ or **	Exponentiation
x %% y	Modulo (remainder of integer division)
x %/% y	Integer division
==	Equal to
!=	Differs from
>	Greater than
<	Less than
>=	Greater than or equal to
<=	Less than or equal to
&	Logical and
|	Logical or
!	Logical not

R is designed for statistical analysis, and therefore it provides many built-in mathematical functions. Some are listed in the next box.

Function	Description
abs(x)	Absolute value
sqrt(x)	Square root
ceiling(x)	Nearest integer $\geq x$
floor(x)	Nearest integer $\leq x$
trunc(x)	Integer part
round(x, digits = n)	Round x to n digits
cos(x), sin(x), tan(x), etc.	Trigonometric functions
log(x)	Natural logarithm
log10(x)	Base 10 logarithm
exp(x)	e^x

Let's have a closer look at the logical operators. They create a logical value by comparing two elements:

```
> 5 > 3
[1] TRUE
> 5 == (10 / 2)
```

```
[1] TRUE
> 6 > 2 ^ 4
[1] FALSE
> 6 >= (2 * 3)
[1] TRUE
> (5 > 3) & (7 < 5) # logical AND
[1] FALSE
> (5 > 3) | (7 < 5) # logical OR
[1] TRUE
```

We can use the operator %in% to search for matches. It returns a logical value (i.e., TRUE or FALSE) indicating whether or not a match exists.

```
> 3 %in% 1:5 # 1:5 generates a sequence
[1] TRUE
# c() combines values into a vector or list
# all arguments are converted to the same type
> c(2, "good to be", TRUE)
[1] "2"          "good to be" "TRUE"
> 2 %in% c(2, "good to be", TRUE)
[1] TRUE
# test for multiple elements at once
> 1:8 %in% c(1, 3, 5)
[1]  TRUE FALSE  TRUE FALSE  TRUE FALSE FALSE FALSE
```

8.7 Assignment and Data Types

When programming in R, you assign values to *variables*: a variable is a "box" that can contain a value or object. The assignment command in R is <- (less-than sign, followed by a minus sign[6]); using the equals sign (=) for assignment is allowed, but deprecated.

```
> x <- 5 # assign the value 5 to x
> x * 2 # use variable x to perform operations
[1] 10
```

---

6. In RStudio you can use Alt+− (i.e., press Alt and then the minus sign).

```
> x <- 7 # assign new value to variable
> x * 2
[1] 14
```

Note that the previous value is overwritten whenever a new value is assigned to the variable. To list all the variables you have created, type `ls()`. To remove a variable x from the current R session, type `rm(x)`.

R can handle different types of data. In the following, we assign a new value to the variable x, each time altering its *data type*. To determine the type of variable x, use the command `class`. You can also test whether a variable is of a certain type by using the functions `is.numeric`, `is.character`, etc.

```
# integer (natural numbers)
> x <- as.integer(5)
> class(x)
[1] "integer"
# numeric (real numbers)
> x <- pi
> class(x)
[1] "numeric"
# complex (complex number)
> x <- 1 + 3i
> class(x)
[1] "complex"
# logical (TRUE or FALSE)
> x <- (5 > 7)
> class(x)
[1] "logical"
# character (strings)
> x <- "hello"
> class(x)
[1] "character"
> is.numeric(x)
[1] FALSE
```

If you need to change the type of a variable (an operation called *casting*), use the functions `as.character`, `as.integer`, `as.numeric`, etc. In RStudio, if you type the first few letters of the name of a function or a variable, and hit Tab, you can see all the possible ways to complete the name; try typing as. and hit Tab to see all the available functions for type conversion. Some examples:

```
> x <- 5
> as.character(x)
[1] "5"
> as.logical(x) # only 0 is FALSE
[1] TRUE
> y <- "07.123" # assign a character string
> x < y # beware: comparing different types
[1] FALSE
> x < as.numeric(y) # cast string to numeric
[1] TRUE
```

8.8 Data Structures

R ships with several *data structures*, which can be used to organize your data. Knowing the characteristics and specific operations of each data structure allows you to write better and more compact code.

8.8.1 Vectors

The most basic data structure in R is the vector, which is an ordered collection of values of the same type. Vectors can be created by concatenating different values with the command c:

```
> x <- c(2, 3, 5, 27, 31, 13, 17, 19)
> x
[1]  2  3  5  27 31 13 17 19
```

You can access the elements of a vector by their index. In contrast to Python, R is 1-indexed, meaning the first element is indexed at 1, the second at 2, etc. You can access the elements of a vector by specifying their positions:

```
> x[3]
[1] 5
> x[8]
```

```
[1] 19
> x[9] # what if the element does not exist?
[1] NA
```

You can extract several elements at once (i.e., create another vector) using the colon (:) command, or by concatenating the indices:

```
> x[1:3]
[1] 2 3 5
> x[4:7]
[1]  27 31 13 17
> x[c(1,3,5)]
[1]  2  5 31
```

Given that R was born for statistics, there are many statistical functions you can perform on vectors:

```
> length(x)
[1] 8
> min(x)
[1] 2
> max(x)
[1] 31
> sum(x) # sum all elements
[1] 117
> prod(x) # multiply all elements
[1] 105436890
> median(x) # median value
[1] 15
> mean(x) # arithmetic mean
[1] 14.625
> var(x) # unbiased sample variance
[1] 119.4107
> mean(x ^ 2) - mean(x) ^ 2 # population variance
[1] 104.4844
> summary(x) # print a summary
   Min. 1st Qu.  Median    Mean 3rd Qu.    Max.
   2.00    4.50   15.00   14.62   21.00   31.00
```

You can generate vectors of sequential numbers using the colon command:

```
> x <- 1:10
> x
[1]  1  2  3  4  5  6  7  8  9 10
```

For more complex sequences, use seq:

```
> seq(from = 1, to = 5, by = 0.5)
[1] 1.0 1.5 2.0 2.5 3.0 3.5 4.0 4.5 5.0
```

To repeat a value or a sequence several times, use rep:

```
> rep("abc", 3)
[1] "abc" "abc" "abc"
> rep(c(1,2,3), 3)
[1] 1 2 3 1 2 3 1 2 3
```

Intermezzo 8.1

(a) Create a vector containing all the even numbers between 2 and 100 (inclusive) and store it in the variable z.
(b) Extract all the elements of z that are divisible by 12. How many elements match this criterion?
(c) What is the sum of all the elements of z?
(d) Is the sum of the elements of z equal to 51×50?
(e) What is the product of elements 5, 10, and 15 of z?
(f) Create a vector y that contains all numbers between 0 and 30 that are divisible by 3. Find the five elements of y that are also elements of z.
(g) Does seq(2, 100, by = 2) produce the same vector as (1:50) * 2?
(h) What happens if you type z ^ 2?

8.8.2 Matrices

A matrix is a two-dimensional table of values. In the case of numeric values, you can perform common matrix operations (e.g., product, inverse, decomposition):

```
# create matrix
# indicate values, number of rows, number of columns
> A <- matrix(c(1, 2, 3, 4), 2, 2)
> A
     [,1] [,2]
[1,]    1    3
[2,]    2    4
> A %*% A # matrix product
     [,1] [,2]
[1,]    7   15
[2,]   10   22
> solve(A) # matrix inverse
     [,1] [,2]
[1,]   -2  1.5
[2,]    1 -0.5
> A %*% solve(A)
     [,1] [,2]
[1,]    1    0
[2,]    0    1
> diag(A) # vector containing the diagonal elements
[1] 1 4
> B <- matrix(1, 3, 2) # 3 x 2 matrix filled with 1s
> B
     [,1] [,2]
[1,]    1    1
[2,]    1    1
[3,]    1    1
> B %*% t(B) # transpose
     [,1] [,2] [,3]
[1,]    2    2    2
[2,]    2    2    2
[3,]    2    2    2
# by default, matrices are filled by column
> Z <- matrix(1:9, 3, 3)
> Z
     [,1] [,2] [,3]
[1,]    1    4    7
[2,]    2    5    8
[3,]    3    6    9
# to fill by rows specify matrix(1:9, 3, 3, byrow = TRUE)
```

Determine the dimensions of a matrix (the numbers of rows and columns) using the following commands:

```
> dim(B)
[1] 3 2
> nrow(B)
[1] 3
> ncol(B)
[1] 2
```

Use indices to access a particular row/column of a matrix:

```
> Z
     [,1] [,2] [,3]
[1,]   1    4    7
[2,]   2    5    8
[3,]   3    6    9
> Z[1, ] # first row
[1] 1 4 7
> Z[, 2] # second column
[1] 4 5 6
# select submatrix with coefficients of second
# and third columns of the first two rows
> Z[1:2, 2:3]
     [,1] [,2]
[1,]   4    7
[2,]   5    8
# concatenate positions to index nonadjacent rows/columns
> Z[c(1,3), c(1,3)]
     [,1] [,2]
[1,]   1    7
[2,]   3    9
```

Most functions consider all coefficients in the matrix:

```
> sum(Z)
[1] 45
> mean(Z)
[1] 5
```

Arrays

If you need tables with more than two dimensions (i.e., tensors), use arrays:

```
# create array
# specify dimensions (rows, columns, matrices)
> M <- array(1:24, dim = c(4, 3, 2))
> M
, , 1

     [,1] [,2] [,3]
[1,]   1    5    9
[2,]   2    6   10
[3,]   3    7   11
[4,]   4    8   12


, , 2

     [,1] [,2] [,3]
[1,]  13   17   21
[2,]  14   18   22
[3,]  15   19   23
[4,]  16   20   24
# determine dimensions
> dim(M)
[1] 4 3 2
```

You can access the elements as for matrices. One thing you should pay attention to is that R drops dimensions that are not needed. So, if you access a "slice" of a three-dimensional array,

```
> M[,,1]
     [,1] [,2] [,3]
[1,]   1    5    9
[2,]   2    6   10
[3,]   3    7   11
[4,]   4    8   12
```

you obtain a matrix:

```
> dim(M[,,1])
[1] 4 3
```

To avoid this behavior, add drop = FALSE when subsetting:

```
> dim(M[,,1, drop = FALSE])
[1] 4 3 1
```

8.8.3 Lists

Vectors are a good choice when each element is of the same type (e.g., numbers, strings). Lists are used when we want to store elements of different types, or more complex objects (e.g., vectors, matrices, even lists of lists). Each element of the list can be referenced either by its index, or by a label:

```
# create list containing two vectors
> mylist <- list(Names = c("a", "b", "c", "d"),
                 Values = c(1, 2, 3))
> mylist
$Names
[1] "a" "b" "c" "d"

$Values
[1] 1 2 3
> mylist[[1]] # access first element within list by index
[1] "a" "b" "c" "d"
> mylist[[2]] # access second element by index
[1] 1 2 3
> mylist$Names # access element by label
[1] "a" "b" "c" "d"
> mylist[["Names"]] # another way to access by label
```

```
[1] "a" "b" "c" "d"
> mylist[["Values"]][3] # third element in vector
[1] 3
```

8.8.4 Strings

If you are performing extensive and complex text manipulation or analysis, Python might be a better choice (see chapter 3); however, R has several built-in functions to manipulate text.

```
# create a string
> x <- "Sample-36"
# split string on specific character
> strsplit(x, '-')
[[1]]
[1] "Sample" "36"
# format text and variables using placeholders
# check documentation for available data types
> sprintf("%s contains %d nucleotide types", "DNA", 4)
[1] "DNA contains 4 nucleotide types"
# extract a substring
> substr(x, start = 8, stop = 9)
[1] "36"
# replace substring
> sub("36", "39", x)
[1] "Sample-39"
# join strings and insert separator
> paste(x, "is significantly smaller", sep = " ")
[1] "Sample-36 is significantly smaller"
> nchar(x) # returns the number of characters
[1] 9
> toupper(x) # set to uppercase
[1] "SAMPLE-36"
> tolower(x) # set to lowercase
[1] "sample-36"
```

8.8.5 Data Frames

Data frames contain data organized like in a spreadsheet. The columns (typically representing different measurements) can be of different types (e.g., one column could be the date of measurement, another the weight of the individual, or the volume of the cell, or the treatment of the sample), while the rows typically represent different samples.

When you read a spreadsheet file in R, it is automatically stored as a data frame. The difference between a matrix and a data frame is that in a matrix all the values are of the same type (e.g., all numeric), while in a data frame each column can be of a different type.

Because typing a data frame by hand would be tedious, let's use a data set that is already available in R:

```
> data(trees) # load data set "trees"
> str(trees) # structure of data frame
'data.frame':   31 obs. of  3 variables:
 $ Girth : num  8.3 8.6 8.8 10.5 10.7 ...
 $ Height: num  70 65 63 72 81 ...
 $ Volume: num  10.3 10.3 10.2 16.4 18.8 ...
> ncol(trees) # number of columns
[1] 3
> nrow(trees) # number of rows
[1] 31
> head(trees) # print the first few rows
  Girth Height Volume
1   8.3     70   10.3
2   8.6     65   10.3
3   8.8     63   10.2
4  10.5     72   16.4
5  10.7     81   18.8
6  10.8     83   19.7
> trees$Girth # select column by name
 [1]  8.3  8.6  8.8 10.5 10.7 ...
# select column by name; return first 5 elements
> trees$Height[1:5]
[1] 70 65 63 72 81
```

```
> trees[1:3, ] # select rows 1 through 3
  Girth Height Volume
1  8.3     70   10.3
2  8.6     65   10.3
3  8.8     63   10.2
# select rows 1 through 3; return column Volume
> trees[1:3, ]$Volume
[1] 10.3 10.3 10.2
> trees <- rbind(trees, c(13.25, 76, 30.17)) # add a row
> trees_double <- cbind(trees, trees) # combine columns
# change column names
> colnames(trees) <- c("girth", "height", "volume")
```

Intermezzo 8.2

   (a) What is the average height of the cherry trees?

   (b) What is the average girth of those that are more than 75 ft tall?

   (c) What is the maximum height of trees with a volume between 15 and 35 ft^3?

8.9 Reading and Writing Data

In most cases, you will not generate your data in R, but import it from a file. By far the best option is to have your data in a comma-separated value text file or in a tab-separated file. Then, you can use the function read.csv (or read.table) to import your data.

```
# to read a CSV file named MyFile.csv
> read.csv("MyFile.csv")
# the file contains a header
> read.csv("MyFile.csv", header = TRUE)
# specify column separator
> read.csv("MyFile.csv", sep = ';')
# skip first 5 lines
> read.csv("MyFile.csv", skip = 5)
```

Note that columns containing strings are typically converted to "factors" (categorical values, useful when performing regressions). To avoid this behavior, you can specify stringsAsFactors = FALSE when calling the function.

Similarly, you can save your data frames using write.table or write.csv. Suppose you want to save the data frame MyDF:

```
# write data to CSV file
> write.csv(MyDF, "MyFile.csv")
# append to end of file
> write.csv("MyFile.csv", append = TRUE)
# include row names
> read.csv("MyFile.csv", row.names = TRUE)
# exclude column names
> read.csv("MyFile.csv", col.names = FALSE)
```

Let's look at an example: Read a file containing data on the sixth chromosome for a number of Europeans[7] (make sure you're in the sandbox directory first!):

```
> ch6 <- read.table("../data/H938_Euro_chr6.geno", header
    ↳ = TRUE)
```

Note that header = TRUE means that we want to take the first line to be a header containing the column names. How big is this table?

```
> dim(ch6)
[1] 43141      7
```

We have 7 columns, but more than 40K rows! Let's see the first few:

```
> head(ch6)
  CHR        SNP A1 A2 nA1A1 nA1A2 nA2A2
1   6  rs4959515  A  G     0    17   107
2   6   rs719065  A  G     0    26    98
```

---

7. Data adapted from Stanford HGDP SNP Genotyping Data hagsc.org/hgdp/ by John Novembre.

```
3   6   rs6596790   C   T       0       4   119
4   6   rs6596796   A   G       0      22   102
5   6   rs1535053   G   A       5      39    80
6   6  rs12660307   C   T       0       3   121
```

and the last few:

```
> tail(ch6)
        CHR         SNP A1 A2 nA1A1 nA1A2 nA2A2
43136     6  rs10946282  C  T     0    16   108
43137     6   rs3734763  C  T    19    56    48
43138     6    rs960744  T  C    32    60    32
43139     6   rs4428484  A  G     1    11   112
43140     6   rs7775031  T  C    26    56    42
43141     6  rs12213906  C  T     1    11   112
```

The data contain the number of homozygotes (nA1A1, nA2A2) and heterozygotes (nA1A2) for 43,141 single nucleotide polymorphisms (SNPs) obtained by sequencing European individuals:

CHR   The chromosome (6 in this case)
SNP   The identifier of the single nucleotide polymorphism
A1   One of the alleles
A2   The other allele
nA1A1   The number of individuals with the particular combination of alleles

Intermezzo 8.3

(a) How many individuals were sampled? Find the maximum of the sum nA1A1 + nA1A2 + nA2A2. Note: you can access the columns by index (e.g., ch6[,5]) or by name (e.g., ch6$nA1A1 or also ch6[,"nA1A1"]).

(b) Try using the function rowSums to obtain the same result.

(c) For how many SNPs are all sampled individuals are homozygous (i.e., all A1A1 or all A2A2)?

(d) For how many SNPs are more than 99% of the sampled individuals homozygous?

8.10 Statistical Computing Using Scripts

Now that we are more familiar with the basics of R, we turn to writing programs. Typically, you will write more complex programs in a text file (called a script), with extension .R. Before writing our first programs, this section will demonstrate how to organize your code to make it readable and easy to understand.

8.10.1 Why Write a Script?

You could actually accomplish almost everything you need for your research without writing any scripts—simply type the commands in the console one at a time. However, by organizing your work into well-documented scripts, you can

recycle: You will encounter similar problems in the future, and, having a script, you will be almost done before you even start.

automate: You will need to repeat the analysis on a different data set, or slightly tweak it in response to comments. This will be quick if you have a script, while repeating the analysis from scratch would take considerably longer (not to mention the possibility of making mistakes).

document: By organizing your code in a script, you will know exactly what you did to obtain your results. You will much appreciate this precise documentation when you write the "Methods" section of your paper or thesis.

share: Having a script makes it much easier to share your analysis with other scientists. You can ask your coauthors to examine your code for errors before you publish and eventually share it with other scientists who want to conduct their analysis exactly as described in your "Methods" section.

8.10.2 Writing Good Code

Before we start writing scripts, we have a few suggestions on how to organize and format your code:

Names

Use descriptive file names for your scripts; use underscore (_) to separate words.

```
1   # good file names
    plotting.R
    model_fitting.R
4
    # bad file names
    misc.R
7   stuff.r
    load_me.R
```

Likewise, choose informative names for your objects and functions:

```
1   # good object names (use nouns!)
    my_variance
    radius
4   body_mass

    # bad object names
7   tmp5
    foo
    good
10
    # to avoid confusion, never call your objects
    I # uppercase i
13  O # uppercase o
    o # lowercase O
    l # lowercase L
16
    # good function names (use verbs!)
    calculate_cv
19  read_fasta

    # bad (i.e., uninformative) function names
```

```
22    f1
      faster_version
      this_works_well
```

Spacing and Parentheses

Put a space before and after an operator (the only exception is a colon), and before a left parenthesis (function arguments are the only exception). Add a space after each comma, but not before. A white line separates functions. Use parentheses to make complex calculations easier to understand.

```
      # good
      x <- 5 * 7
3     y <- 7 * (x ^ 2)
      m <- matrix(25, 5, 5)
      z <- mean(x, na.rm == TRUE)
6     i <- t + 1
      z <- (x * y) + (x2 * y2)

9     if (b == 5) {
        do(something)
      } else {
12      do(something_else)
      }

15    # bad
      z = "bad_assignment_style"
      x<-5*7
18    y <- 7*x^2
      m <- matrix(25 , 5 , 5)
```

While it is good practice to write well-formatted code in the first place, you can also use the reformatting function in RStudio: mark your code and select "Reformat Code" from the "Code" menu. Alternatively, you can use the package formatR.

From now on, we will write scripts, and save them into the sandbox directory within the CSB/r/ directory. In this book, scripts are printed with a lighter gray background and line numbers.

8.11 The Flow of the Program

When you execute an R script, R reads the lines of code in order from the top to the bottom. Every time R encounters a command, it will execute it. Thus, in its simplest form, an R program is simply a sequence of commands.

However, it is often essential to modify this linear flow of the code: you might have commands that need to be run only if certain conditions are met, commands that need to be run over several files/data sets, commands that you repeat several times, etc. In this section, we show how to modify the flow of a program.

8.11.1 Branching

The simplest modification of the linear flow of a program is given by conditional branching: if a certain condition is met, then certain commands are executed; otherwise, other commands may be executed.

Let's create a new script by pressing Ctrl+Shift+N. Save the script (Ctrl+S) as `conditional.R` in the `CSB/r/sandbox`. Make sure to set the working directory to the `sandbox`. In RStudio, you can choose the working directory by pressing Ctrl+Shift+H. Alternatively, use the `setwd()` command. Now start your script by typing

```
z <- readline(prompt = "Enter a number: ")
```

The function `readline` reads input from the user. It returns a string. Let's convert the string to numeric type:

```
# request user input and store in variable
z <- readline(prompt = "Enter a number: ")
# convert to numeric variable
z <- as.numeric(z)
```

Now we want to determine whether the number is even or odd, and print a statement. If the remainder on division by 2 equals 0 (z %% 2 == 0), then the number z is even.

```
     z <- readline(prompt = "Enter a number: ")
2    z <- as.numeric(z)

     # test whether remainder equals 0
5    # if yes, print even statement
     # else, print odd statement
     if (z %% 2 == 0) {
8    # concatenate and print statement
       print(paste(z, "is even"))
     } else {
11   # print alternative statement
       print(paste(z, "is odd"))
     }
```

The anatomy of the if statement is

```
1    if (a condition is TRUE) {
       execute these commands
     } else {
4      execute these other commands [optional]
     }
```

We switch to the RStudio console and run the script a few times by pressing Ctrl+Shift+S:

```
Enter a number: 12
[1] "12 is even"
[...]
```

If you are not working in RStudio or want to execute a script that is not currently open in RStudio, you can source it:

```
# specify a path to the script and execute it
> source("conditional.R")
Enter a number: 22
```

```
[1] "22 is even"
> source("conditional.R")
Enter a number: 27
[1] "27 is odd"
```

Intermezzo 8.4

Add code to the script so that

(a) if $z > 100$, the program prints z^3;
(b) if z is divisible by 17, the program prints $\sqrt{z}$;
(c) if $z < 10$, the program prints a vector containing the numbers between 1 and z.

8.11.2 Loops

Another way to modify the flow of a program is to write a loop. A loop is a series of commands that are repeated a number of times. For example, you want to run the same analysis on different data sets that you have collected, or you want to plot the results contained in a set of files, or you want to test your simulation over a number of parameter sets, etc.

R provides you with two ways to loop over blocks of commands: the `for` loop, and the `while` loop. Let's start with the `for` loop, which is used to iterate over a vector (or a list): for each value of the vector, a series of commands will be run, as shown by the following example, which you can type in a script called `forloop.R`:

```
myvec <- 1:10 # vector with numbers from 1 to 10

for (i in myvec) {
  a <- i ^ 2
  print(a)
}
```

In the code above, the variable `i` takes the value of each element of `myvec` in sequence. Inside the block defined by the `for` loop, you can use the variable `i` to perform operations. Note that `i` is only an example of a variable name.

This is the anatomy of a `for` statement:

```
for (variable in list_or_vector) {
    execute these commands
} # automatically moves to the next value
```
3

A `for` loop is used when you know that you want to perform the analysis using a given set of values (e.g., run over all files in a directory, all samples in your data, all sequences of a FASTA file).

The `while` loop is used when the commands need to be repeated while a certain condition is true, as shown by the following example, which you can type in a script called `whileloop.R`:

```
i <- 1

while (i <= 10) {
    a <- i ^ 2
    print(a)
    i <- i + 1
}
```
3

6

The script performs exactly the same operations we wrote for the `for` loop above. Note that you need to update the value of i (using i <- i + 1), otherwise the loop will run forever (infinite loop—to terminate click on the "Stop" button in the top-right corner of RStudio). The anatomy of the `while` statement is

```
while (condition is TRUE) {
    execute these commands
    update the condition
}
```
2

You can exit a loop by using the command `break`. For example,

```
i <- 1

while (i <= 10) {
    if (i > 5) {
        break
```
2

5

```
      }
      a <- i ^ 2
8     print(a)
      i <- i + 1
    }
```

Intermezzo 8.5

What do these codes do? Try to guess what each loop does, and then create and run a script to confirm your intuition.

1. Code:

```
    z <- seq(1, 1000, by = 3)
2   for (k in z) {
      if (k %% 4 == 0) {
        print(k)
5     }
    }
```

2. Code:

```
    z <- readline(prompt = "Enter a number: ")
    z <- as.numeric(z)
3
    isthisspecial <- TRUE
    i <- 2
6   while (i < z) {
      if (z %% i == 0) {
        isthisspecial <- FALSE
9       break
      }
      i <- i + 1
12  }

    if (isthisspecial == TRUE) {
15    print(z)
    }
```

8.12 Functions

So far, we have used many *built-in functions* (e.g., length, dim). What makes R powerful is the ability to define your own functions and invoke them within your programs. These are called *user-defined functions*.

Here is the general anatomy of a function:

```
my_func_name <- function([optional arguments]) {
    operations
    return(value) [optional]
}
```

Each function needs a function name, which you can use to invoke the function in your script. You then type the keyword function, with optional arguments in parentheses. The code of the function is contained between the curly brackets. At the end, a value may be returned. We will later examine functions that do not return a value.

Here is an example of a function. We want to determine whether a number is "triangular." Triangular numbers count objects that can be arranged in an equilateral triangle. For example, 1, 3, 6, 10, and 15 are triangular numbers:

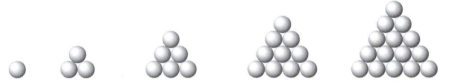

Each triangular number T can be written as $T = n(n + 1)/2$ (e.g., $T = 1$, $n = 1$; $T = 3, n = 2$). Hence, an integer y is triangular if

$$n = (\sqrt{8y + 1} - 1)/2 \tag{8.1}$$

is also an integer. We can write a function that checks whether an integer is triangular:

```
# check whether a number is triangular
# take a single argument, y
is_triangular <- function(y){
    n <- (sqrt((8 * y) + 1) - 1) / 2
    # if triangular, then n should be an integer
    if (as.integer(n) == n) {
```

```
       # if condition is true, then return TRUE
 8     return(TRUE)
     }
       # if condition is not true (i.e., n is not an integer),
          return FALSE
11   return(FALSE)
   }
```

Save this function in the script `triangular.R`. In the console, we can now check that everything works as expected:

```
> source("triangular.R") # read the script
> is_triangular(91)
[1] TRUE
> is_triangular(78)
[1] TRUE
> is_triangular(4)
[1] FALSE
> is_triangular(56)
[1] FALSE
```

Now let's add another function to the script. The second function shall return all the triangular numbers between 1 and `max_number`:

```
       # function to find and store triangular numbers
       # argument is one number (max_number)
 3     # test for triangular numbers from 1 up to max_number
     find_triangular <- function(max_number){
       # create vector with all numbers to be tested
 6     to_test <- 1:max_number
       # create empty vector for storage
     triangular_numbers <- numeric(0)
 9     # iterate using a for loop
     for (i in to_test) {
       # call function to test whether number is triangular
12     if (is_triangular(i)) {
         # if is_triangular returns TRUE, add to result vector
```

```
          triangular_numbers <- c(triangular_numbers, i)
15        }
      }
      # when all numbers are tested (i.e., loop exited)
18    # print results to screen
      print(paste("There are", length(triangular_numbers),
                  "triangular numbers between 1 and ",
                     max_number))
21    return(triangular_numbers)
  }
```

Save the script and start testing in the console:

```
# load the script again to make both functions accessible
> source("triangular.R")
> find_triangular(100)
[1] "There are 13 triangular numbers between 1 and  100"
[1]  1  3  6 10 15 21 28 36 45 55 66 78 91
> find_triangular(1000)
[1] "There are 44 triangular numbers between 1 and  1000"
[1]   1   3   6   10   15   21   28   36   45   55   66   78  ...
> find_triangular(10000)
[1] "There are 140 triangular numbers between 1 and 10000"
  [1]   1   3   6   10   15   21   28   36   45  ...
```

Here we wrote two functions, each taking an argument (y for is_triangular; max_number for find_triangular). You can also write functions that do not require arguments. When you need several arguments, separate them using commas. If you want to have a *default value* (i.e., a value that is used if the user does not specify one) for one or more arguments, you can set them in the function declaration using the equals sign. Each function can return only *one* value. If you need to return multiple variables, organize them in a list or vector and return it.

```
1   # a function with no arguments and returning no value
    tell_fortune <- function() {
      # runif(1) generates a random number between 0 and 1
```

```
 4    if (runif(1) < 0.3) {
        print("Today is going to be a great day for you!")
      } else {
 7      print("You should have stayed in bed")
      }
    }

10
    # a function taking multiple values
    # and returning a vector
13  order_three <- function(x, y, z) {
      return(sort(c(x, y, z)))
    }

16
    # a function taking multiple values
    # and returning a list
19  order_three_list <- function(x, y, z) {
      my_ord <- sort(c(x, y, z))
      return(list("First" = my_ord[1],
22            "Second" = my_ord[2],
              "Third" = my_ord[3]))
    }

25
    # a function making use of default values
    split_string <- function(s, separator = "_") {
28    # if not specified, split using underscore
      return(strsplit(s, separator)[[1]])
    }
```

Note that variables that are specified within a function are *local* and not available outside the function. You can read more on *variable scope* in our explanations using Python in section 4.9.4.

8.13 Importing Libraries

In addition to your own and the many built-in functions, you can also take advantage of the many packages that are available for R. A package is loosely defined as a collection of code, data, documentation, and tests. Each package can, in turn, contain a large number of functions. You can install packages in RStudio using the "Packages" tab located in the lower-right panel. Once a package is installed, you need to load it by typing

```
library([NAME_OF_PACKAGE])
```

R users have currently contributed nearly 12,000 packages to the Comprehensive R Archive Network. Consult section 8.21 to learn how to find packages that are relevant to your research interests.

8.14 Random Numbers

R can sample (pseudo)random numbers from many distributions. This is very useful for simulations! For example,

```
# extract 3 numbers from uniform distribution U[0,1]
> runif(3)
[1] 0.8252214 0.8811069 0.8099231
# sample 4 numbers from normal distribution
# with mean 1 and standard deviation 5
> rnorm(4, mean = 1, sd = 5)
[1] 7.2729430 0.3416511 5.5701139 8.5061745
# draw from Poisson distribution with mean lambda
> rpois(4, lambda = 5)
[1] 4 7 4 7
```

You can easily sample with or without replacement from a vector:

```
> v <- 1:10
> sample(v, 2) # sample without replacement from v
[1] 5 8
> sample(v, 11, replace = TRUE) # sample with replacement
 [1]  9  4  8  8  2  4  6  1  7  8 10
```

Intermezzo 8.6

(a) Add a new column Exp (for "Experiment") to the data set trees and randomly assign a 1 or 2 (with equal probability) to each entry of the data frame.

(b) Write a function that randomly assigns 1 or 2 to column Exp. Use this function in a loop that runs 100 t-tests[a] of tree volume in experiment 1 versus experiment 2. What proportion of tests return "significant" differences when using $p < 0.05$ as a cutoff?

(c) Write a function that takes three arguments, x1, x2, and x3, and determines whether their sum is a pentagonal number. Pentagonal numbers are integers that can be written as $P = n(3n - 1)/2$. Thus, the integer y is pentagonal if $x = (\sqrt{24y + 1} + 1)/6$ is also an integer. For example, 1, 5, 12, 22, and 35 are the first few pentagonal numbers.

---

a. To perform a t-test, see the documentation for the function t.test.

8.15 Vectorize It!

R is very slow at running cycles (e.g., for and while loops). This is because R is a "nimble" language: when you execute a script, R does not know what you are going to perform next, and takes one command at a time. Code in languages such as C is typically compiled before execution, so that at run time, the exact flow of the program is clear, and the program "knows" which variables are present and what their type is. As a metaphor, C is a musician playing a score they have seen before, optimizing each passage, while R is playing it *a prima vista* (i.e., at first sight).

In practical terms, it is often easier to use a for loop. However, if the running time is not satisfactory, you can optimize the code.

R provides many functions that automatically operate elementwise on a vector, which in many cases can remove the need for cycles. Given that vectors contain only one type of data, R needs to figure out the type only once—and not individually for each element—making the code run faster.

In the following example, we want to calculate the means of rows of a matrix. First, we time the processing of sample data when we use a function that relies on a loop. Then we repeat the calculation using a vectorized approach:

```
# generate sample data
# draw 10,000 x 10,000 values from uniform distribution
# fill a matrix of 10,000 cols and 10,000 rows with these
> M <- matrix(runif(10000 * 10000), 10000, 10000)

# function that calculates row means of a matrix
> get_row_means <- function(M) {
+     # set up vector to capture results
+     my_row_means <- rep(0, nrow(M))
+     # loop over rows in matrix
+     for (i in 1:nrow(M)) {
+         # add result to result vector
+         my_row_means[i] <- mean(M[i,])
+     }
+     return(my_row_means)
+ }

# time user-defined function that uses a loop
> system.time(get_row_means(M))
  user   system elapsed
 2.292   0.241   2.651

# time execution of built-in function
# that operates rowwise (i.e., vectorized, without a loop)
> system.time(rowMeans(M))
  user   system elapsed
 0.260   0.008   0.269
```

R has functions that can operate on entire vectors, matrices, and lists. In particular, if you can structure your code so that you are operating on elements of a list, it can save you a lot of time. Here are some useful functions:

```
# lapply: applying the same function to elements of a list
# the result is returned as a list

# example: create a list of matrices
# first, create an empty list
```

3

```
 6   MList <- as.list(rep(NA, length = 20))
     # second, a function to generate small random matrices
     randmat <- function(x) {
 9       return(matrix(rnorm(25), 5, 5))
     }
     # lapply requires two arguments:
12   # the list to use (MList)
     # and the function to apply (randmat)
     MList <- lapply(MList, randmat)

15
     # we can also define a function directly in the lapply call
     # example: find the largest eigenvalue of each matrix
18   Meig <- lapply(MList, function(x)
         return(eigen(x, only.values = TRUE)$values[1]))
     # use unlist() to convert a list into a vector
21   print(unlist(Meig))


     # sapply: applying the same function to elements of a list
24   # the result is returned as a vector


     # example: count occurrences of nucleotides in a sequence
27   # create a list of DNA sequences
     DNAlist <- list(A = 'GTTTCG',
                     B = 'GCCGCA',
30                   C = 'TTATAG',
                     D = 'CGACGA')


33   # write a function that counts nucleotide occurrences
     count_nucl <- function(seq, nucl) {
         # gregexpr returns a list
36       # with starting positions of matches
         pos <- gregexpr(pattern = nucl, text = seq)[[1]]
         # by default, gregexpr returns -1 if there is no match
39       # we modify this behavior so our function returns 0
         if(pos[1] == "-1") {
             return(0)
42       } else {
             return(length(pos))
```

```
45        }
    }

48    # apply the function count_nucl to all elements in DNAlist
      # sapply requires two arguments:
      # the list to use (DNAlist)
51    # and the function to apply (count_nucl)
      # sapply returns a vector
      numAs <- sapply(DNAlist, count_nucl, nucl = 'A')
54    print(numAs)

      numGs <- sapply(DNAlist, count_nucl, nucl = 'G')
57    print(numGs)
```

8.16 Debugging

Errors in the code ("bugs") can be difficult to pinpoint. While the most blatant errors will cause the code to terminate abruptly—and thus are easy to identify—the most pernicious bugs are those that are "silent," so that the code will run, but will not produce the desired result. These can cause great harm to your scientific projects.

Most scientists debug their code by adding a bunch of print statements here and there, trying to zero in on the problem. This is inefficient and unnecessary. Simply add browser() at any point of your code. When R encounters this command, it will enter debugging mode.

For example, create a script debug.R:

```
1    myfun <- function(i, x) {
      for (z in 1:i) {
        x <- x * 2
4       browser()
      }
      return(x)
7    }
```

When we run the script, we enter debugging mode (note the changes in your RStudio console):[8]

```
> myfun(5, 2)    # run the function
Called from: myfun(5, 2)
# see local variables of function
Browse[1]> ls()
[1] "i" "x" "z"
# determine current value of individual variables
Browse[1]> i
[1] 5
Browse[1]> x
[1] 4
Browse[1]> z
[1] 1
# press n to execute next statement
Browse[1]> n
debug at debug.R#3: x <- x * 2
# the hash indicates the number of the line
# see new value of local variables
Browse[2]> x
[1] 4
Browse[2]> n
debug at debug.R#4: browser()
Browse[1]> n
debug at debug.R#3: x <- x * 2
Browse[2]> Q # exit browser and function
# type c to continue executing the function
# but exit the browser
```

8.17 Interfacing with the Operating System

You can call the operating system from within R:

---

8. Naming your variables c, n, or Q is generally a bad idea but especially problematic in debugging mode as these are reserved for function calls.

```
# run a Unix command from within R
> system("wc -l < ../data/H938_Euro_chr6.geno")
43142
```

You can also capture the output from shell commands and save it into R. Everything is treated as text (convert to numeric if necessary):

```
> numlines <- system("wc -l < ../data/H938_Euro_chr6.geno"
    ↳ , intern = TRUE)
> numlines
[1] "43142"
```

You can also use a combination of shell commands and read.table to capture more complex output:

```
> mydf <- read.table(file = textConnection(
                  system("grep 'rs125283' ../data/
                      ↳ H938_Euro_chr6.geno",
                  intern = TRUE)))
> mydf
  V1          V2 V3 V4 V5 V6  V7
1  6 rs12528302  G  A 26 59  39
2  6 rs12528322  G  A  0 21 103
3  6 rs12528313  G  T  1 25  98
4  6 rs12528341  C  T  3 31  90
```

8.18 Running R from the Command Line

So far we have worked in RStudio in "interactive" mode: we type a command, it gets executed, we type another command, and so on. To automate our workflow, run a script on a remote server, or parallelize our work, we need to be able to run an R script from the command line.

You can execute R scripts from the Unix command line by typing

```
$ Rscript my_script_file.R
```

We can additionally pass arguments from the command line to R. This allows us, for instance, to perform the analysis using a specific input file, run an analysis using different parameters, or save a figure using a specific file name. The code at the beginning of the following script shows how this is accomplished:

```
    # get all the command-line arguments
2   args <- commandArgs(TRUE)
    # assign each argument to a variable
    # make sure to convert it to the right type
5   # arguments are string by default

    # check the number of arguments
8   num.args <- length(args)
    print(paste("Number of command-line arguments:", num.args))
    # print all the arguments
11  if (num.args > 0) {
        for (i in 1:num.args) {
            print(paste(i, "->", args[i]))
14      }
    }

17  # we can initially set to default values
    # pay attention to the order
    # optional arguments should be at the end
20  input.file <- "test.txt"
    number.replicates <- 10
    starting.point <- 3.14
23
    if (num.args >= 1) {
        input.file <- args[1]
26  }
    if (num.args >= 2) {
        number.replicates <- as.integer(args[2])
29  }
    if (num.args >= 3) {
        starting.point <- as.double(args[3])
32  }

    print(c(input.file, number.replicates, starting.point))
```

Save this script as `my_script.R` and run it from the command line with different sets of arguments:

```
$ Rscript my_script.R abc.txt 5 100.0
$ Rscript my_script.R abc.txt 5
$ Rscript my_script.R abc.txt
$ Rscript my_script.R
```

Always pay extra attention to the correct order of arguments when executing a script from the command line. As we have seen in the above script, the order determines which variable they get assigned to within R.[9]

8.19 Statistics in R

R ships with many functions for basic statistical analysis, and many more advanced statistical functions can be added by loading additional packages. Here we present some basic commands showing that statistical analysis is easy to perform in R. Please refer to the documentation of the statistical tests, or the resources in section 8.24, to learn more about these commands, their statistical meaning, and interpretation of results.

We will again work with a data set that ships with R:

```
> attach(iris)
> str(iris)
'data.frame':   150 obs. of  5 variables:
 $ Sepal.Length: num  5.1 4.9 4.7 4.6 5 ...
 $ Sepal.Width : num  3.5 3 3.2 3.1 3.6 ...
 $ Petal.Length: num  1.4 1.4 1.3 1.5 1.4 ...
 $ Petal.Width : num  0.2 0.2 0.2 0.2 0.2 ...
 $ Species     : Factor w/ 3 levels "setosa","versicolor"
```

where the command `attach` makes columns accessible without referencing the data set.

---

9. Alternatively, you can name your arguments, so that the order does not matter. This can be accomplished using packages like `argparser` and `optparse`.

```
# instead of iris$Species, we can call Species directly:
> levels(Species)
[1] "setosa"     "versicolor" "virginica"
```

For a first, exploratory analysis, we can invoke the commands summary and table:

```
> summary(iris) # show summary statistics
  Sepal.Length     Sepal.Width     Petal.Length ...
 Min.   :4.300   Min.   :2.000   Min.   :1.000 ...
 1st Qu.:5.100   1st Qu.:2.800   1st Qu.:1.600 ...
 Median :5.800   Median :3.000   Median :4.350 ...
 Mean   :5.843   Mean   :3.057   Mean   :3.758 ...
 3rd Qu.:6.400   3rd Qu.:3.300   3rd Qu.:5.100 ...
 Max.   :7.900   Max.   :4.400   Max.   :6.900 ...


# return a frequency/contingency table with counts
> table(Species)
Species
    setosa versicolor  virginica
        50         50         50


# two-way table allows us to examine relationship
# between categorical variables
> table(Species, Petal.Width)
> table(Species, Petal.Width)
[Output not shown]
```

These commands are two of the many dedicated to exploratory analysis. Before you analyze your data, it is prudent to explore the relationships among your measures, and have a general idea of the variability in the data. Some of the most useful commands are:

- range returns the minimum and maximum in a column (e.g., range(iris$Petal.Width)).
- by produces summaries for data that is divided into groups. For example, by(iris$Sepal.Length, iris$Species, mean) calculates the mean Sepal.Length of each Species.
- cor computes Pearson's correlation between two columns (e.g., cor(iris$Petal.Width, iris$Petal.Length)).
- pairs plots each column against every other column in the data. Useful to see whether certain columns are correlated (e.g., pairs(iris)).

- subset extracts the part of a data.frame that matches a condition. For example, subset(iris, iris$Petal.Width > 0.2) returns the samples with Petal.Width > 0.2.
- rank returns the sample ranks of a column, i.e., the position they would take when sorted from smallest to largest (e.g., rank(iris$Sepal.Length)).
- which.max (or which.min) returns the index of the largest (or smallest) element in a vector (e.g., which.max(iris$Sepal.Length)).

R provides all common statistical tests as built-in functions. For example, let's run a *t*-test. It determines whether the sample means of two groups differ significantly.

```
# t-test on sepal width of two iris species
> t.test(Sepal.Width[Species == "setosa"],
+        Sepal.Width[Species == "versicolor"])

    Welch Two Sample t-test

data:  Sepal.Width[Species == "setosa"] and Sepal.Width[
    ↳ Species == "versicolor"]
t = 9.455, df = 94.698, p-value = 2.484e-15
alternative hypothesis: true difference in means is
not equal to 0
95 percent confidence interval:
 0.5198348 0.7961652
sample estimates:
mean of x mean of y
3.428     2.770
```

We can also easily define a linear regression model. Linear regression describes the relationship between two variables x and y.

```
# linear regression
> linearmod <- lm(Sepal.Width ~ Sepal.Length)
> summary(linearmod)
Call:
lm(formula = Sepal.Width ~ Sepal.Length)
```

```
Residuals:
    Min      1Q  Median      3Q     Max
-1.1095 -0.2454 -0.0167  0.2763  1.3338

Coefficients:
              Estimate Std. Error t value Pr(>|t|)
(Intercept)    3.41895    0.25356   13.48   <2e-16 ***
Sepal.Length  -0.06188    0.04297   -1.44    0.152
---
Signif. codes:  0 '***' 0.001 '**' 0.01 '*' 0.05 '.' 0.1 '
    ↳  ' 1

Residual standard error: 0.4343 on 148 degrees of freedom
Multiple R-squared: 0.01382,    Adjusted R-squared: 0
    ↳ .007159
F-statistic: 2.074 on 1 and 148 DF,  p-value: 0.1519
```

We first construct a linear model (lm) in the form of response variable
(y) ~ predictor (x). We then call the summary function on the output
to retrieve statistical information such as the adjusted R-squared, residual
standard error, and F-statistics.

8.20 Basic Plotting

In R, you can choose different ways to plot your data and results. By default,
R ships with built-in graphical functions. Other functions, or entirely different paradigms, can be enabled using different packages. Here we give a
brief overview of the standard plotting functions that are useful for quick data
exploration. These are also referred to as R "base graphics" as no extra package is required. In the next chapter, we introduce ggplot2, which is a very
powerful and visually appealing alternative for producing figures in R.

   R base graphics are drawn interactively: you can start with a plot, and
then overlay other elements on the existing plot.

8.20.1 Scatter Plots

The function plot is the most basic way to plot points defined by their x- and
y-coordinates.

```
# generate data
> x <- 1:30
> y <- rnorm(30, mean = x)
> y2 <- rnorm(30, mean = x, sd = sqrt(x))
# plot y against x
> plot(y ~ x)
# alternatively
> plot(x, y)
# using lines instead of points
> plot(x, y, type = "l")
# using both
> plot(x, y, type = "b")
# change the type of point
> plot(x, y, type = "b", pch = 4)
# change color
> plot(x, y, type = "b", pch = 4, col = "blue")
# add a line
> abline(c(0, 1)) # intercept, slope
# add another data set
> points(x, y2, col = "red")
# set x-label and y-label
> plot(x, y2, col = "orange", xlab = "my x-label", ylab =
   ↳ "yyy")
# set ranges
> plot(x, y2, xlim = c(1,10))
```

8.20.2 Histograms

The function hist is used to produce simple histograms:

```
# create random data set of 100 numbers
# drawn from Poisson distribution with mean 3
> d1 <- rpois(100, lambda = 3)
# basic histogram
> hist(d1)
# specify desired number of bins
> hist(d1, breaks = 4)
# specify bin edges
```

```
> hist(d1, breaks = c(0, 1, 3, 5, 7, 11, 21))
# use frequencies (default if bins have equal size)
> hist(d1, freq = TRUE)
# use density
> hist(d1, freq = FALSE)
# get the histogram, but without plotting it
> z <- hist(d1, plot = FALSE)
# access elements of the histogram
> z$counts # counts per bin
> z$mids # midpoints of the bins
```

8.20.3 Bar Plots

A bar plot is used to represent data for discrete groups. Here are some examples using the function barplot:

```
> data(islands) # area of islands
> barplot(islands)
> barplot(islands, horiz = TRUE) # horizontal
# change orientation of labels
> barplot(islands, horiz = TRUE, las = 1)
> data(iris)
# color by group level
> barplot(height = iris$Petal.Width, beside = TRUE, col =
    ↳ iris$Species)
```

8.20.4 Box Plots

Box plots are used to show the range of a distribution, and the location of the bulk of its mass. Use the function boxplot:

```
> data(iris)
# set specific colors
> boxplot(iris$Petal.Width ~ iris$Species, col = c("red",
    ↳ "green", "blue"))
```

8.20.5 3D Plotting (in 2D)

To show density plots, or plot a matrix, use `image`:

```
# generate data
> x <- sort(rnorm(100))
> y <- sort(rnorm(50))
> z <- x %o% y # outer product
# 3D plotting in 2D
> image(z) # simple density plot
> filled.contour(z) # fancier density plot
```

8.21 Finding Packages for Biological Research

R is the most popular statistical computing language among biologists due to its highly specialized packages, often written by biologists for biologists.[10] In this chapter, we have explored basic R features, but you can find highly specialized packages to address your research questions. Here are some suggestions for finding an appropriate package:

The Comprehensive R Archive Network (CRAN) offers several ways to find specific packages for your task. You can either browse packages and their short descriptions (computingskillsforbiologists.com/rpackages), or select a scientific field of interest and browse through lists of packages related to that discipline (computingskillsforbiologists.com/rpackagesbyfield).[11] From within your R terminal or RStudio you can also call the function `RSiteSearch`, which submits a search query to the website search.r-project.org. The website rseek.org casts an even wider net, including not only package names and their documentation, but also blogs and mailing lists related to R.

If your research involves the analysis of high-throughput genomic data, you should have a look at the packages provided by Bioconductor (computingskillsforbiologists.com/bioconductor).

---

10. You can contribute a package, too! The RStudio support provides guidance on how to start developing R packages (computingskillsforbiologists.com/developrpackages) and Hadley Wickham's free online book will make you a pro (r-pkgs.had.co.nz).

11. Install all packages related to a field at once by using the `ctv` package. First install the `ctv` package, load it, and type `install.views("genetics")`, or type `update.views("genetics")` to install every available package to conduct statistical genetics.

8.22 Documenting Code

We have discussed extensively the importance of properly documenting your code. R and RStudio provide special packages that make it easy to write incredibly clear and pretty documentation for your code, without the need to put in any extra effort. For example, the package knitr enables you to document your code using Markdown, LaTeX, etc., and either run it as an R script, or dynamically produce reports including the text describing the code, the code itself, and the results!

To give you a taste of how beautiful your code and documentation could look, we provide all solutions for the exercises in chapters 8 and 9 in R Markdown. As an example, open one of the .Rmd files that you find in the solutions directory in RStudio. You can execute the file as a script by hitting Ctrl+Shift+S.

At the beginning of the .Rmd file, you will notice the definition of the output format (e.g., PDF or HTML among many others). You *knit* a file in RStudio to produce the desired output format. Hit Ctrl+Shift+K to start the process (when launched for the first time, RStudio may prompt you to install some packages).

As you see in our solution files, you can document code extensively using plain text. You can alternate documentation and code chunks by using special Markdown tags (as seen in our exercise solutions). Code *chunks* are set within triple tick marks, followed by {r}:

```
   ```{r}
2  Everything between these marks will be
   evaluated in the R console
   ```

5 Text not enclosed in triple tick marks is not evaluated.
   This is the spot to write extensive documentation.

8 You can name your code chunks:
   ```{r histogram}
   hist(iris$Sepal.Width)
11   ```
```
```

The box lists a few examples of R Markdown syntax that might come in handy when documenting your code:

This is a first-level headline
This is a second-level headline

italics

bold

'verbatim'

<http://www.WEBSITE.com>[link](www.WEBSITE.com)

8.23 Exercises

8.23.1 Self-Incompatibility in Plants

Goldberg et al. (2010) studied self-incompatibility in Solanaceae. Self-incompatible plants can recognize and reject their own pollen. The file data/Golberg2010_data.csv contains a list of 356 species, along with a flag determining the self-incompatibility status: 0 stands for self-incompatible, 1 for self-compatible, 2–5 for more complex scenarios.

1. Write a program that counts how many species are in each category of Status. The output should be a data.frame:

```
  Status count
1      0   116
2      1   196
3      2    17
4      3     1
5      4    25
6      5     1
```

2. Write a program that builds a data.frame specifying how many species are in each Status for each genus (note that each species name starts with the genus, followed by an underscore).

8.23.2 Body Mass of Mammals

Smith et al. (2003) compiled a database of the body mass of mammals of the late Quaternary period. Your goal is to calculate the average body mass of the species in each family.

1. Write a script (read_mass.R) that reads and cleans the data:
   - Read the file Smith2003_data.txt; column 7 contains the data on body mass (in log kilograms).
   - The authors used -999 to mark missing data. Use NA instead.
   - Add column names.
2. Write another script (body_mass_family.R) that calls read_mass.R and then calculates the average body mass per family.
3. Using the command

```
> system.time(source("body_mass_family.R"))
```

   measure the time it takes to run the analysis.
4. There are many ways to accomplish this task. Rewrite the analysis using an alternative method, make sure it returns the same results, and time the alternative solution. Which one is more readable? Which one is faster?

8.23.3 Leaf Area Using Image Processing

In this exercise, we will get a glimpse of the image processing capabilities of R. We want to determine the projected leaf area of plants using photos, and analyze whether the leaves have grown significantly over the course of two days. The directory CSB/r/data/leafarea/ contains images of plants at two time points (t1 and t2). The data have been collected by Madlen.

1. Write a for loop that processes all images using the function getArea, which is provided in CSB/r/solutions/getArea.R. The function accepts a single file name as an argument, and returns the projected leaf area, measured in pixels. Your loop should record the leaf area for each image, and store it in the data frame results. To loop over all files, you can use the function list.files along with its pattern matching option, to produce a list of all the files with extension .jpg in the directory SC/r/data/leafarea/. Work in your sandbox or change paths in the getArea.R function accordingly.

2. Plot the area of each plant as measured at time point 1 versus time point 2.

3. Determine whether the plants significantly differ at time points 1 and 2 using a paired *t*-test.

8.23.4 Titles and Citations

Letchford et al. (2015) found an interesting pattern: papers that have shorter titles tend to fare better in terms of citations. They took top-cited papers from a variety of journals, and ranked them by title length (in number of characters), and number citations received (as of November 2014). Then they performed Kendall's τ-test to see whether these rankings are correlated. A negative correlation would mean that articles with longer titles tend to be ranked low for citations.

The file Letchford2015_data.csv contains the data needed to replicate their results.

1. Write a program that performs the test described above using all the papers published in 2010. The program should do the following: (1) read the data; (2) extract all the papers published in 2010; (3) rank the articles by citations and by title length;[12] (4) compute Kendall's τ expressing the correlation between the two rankings.[13] For this data set, the authors got a τ of about −0.07 with a significant *p*-value.

2. Write a function that repeats the analysis for a particular journal–year combination. Try to run the function for the top scientific publications *Nature* and *Science*, and for the top medical journals *The Lancet* and *New Eng J Med* for all years in the data (2007–2013). Do you always find a negative, significant correlation (i.e., negative τ with low *p*-value)?

8.24 References and Reading

Many books and online resources are available to learn R. You will also find introductions targeting specific scientific fields, or particular types of analysis. Finally, many books use R as a springboard to learn about modeling and statistical theory.

---

12. See the documentation of the function rank.
13. See the documentation of the function cor.test.

Books

B. M. Bolker, *Ecological Models and Data in R*, Princeton University Press, 2008.

 A practical introduction to statistical methods for ecology in R.

The *Use R!* series by Springer.

 This series provides more than 50 titles covering both introductory (*A Beginner's Guide to R, R by Example, Introductory Statistics with R*) and specific topics (*Numerical Ecology With R, A Primer of Ecology with R, Nonlinear Regression with R, Analysis of Phylogenetics and Evolution with R*).

K. Soetaert, P. M. J. Herman, *A Practical Guide to Ecological Modeling: Using R as a Simulation Platform*, Springer, 2009.

 Emphasizes the conceptual and mathematical bases of modeling, covering a wide range of ecological models.

Patrick Burns, *The R Inferno*, computingskillsforbiologists.com/rinferno, 2011.

 If you think programming in R is like going through hell, make sure you read this.

Online Resources

The Comprehensive R Archive Network provides source code, packages, and many documents and tutorials:

 cran.r-project.org.

Clear and comprehensive documentation on R:

 statmethods.net.

R for MATLAB users:

 computingskillsforbiologists.com/rformatlabusers.

Software Carpentry workshop on R for Python programmers:

 computingskillsforbiologists.com/introtor.

Learn R in *Y* minutes:

 computingskillsforbiologists.com/learnxiny.

Slides on the similarities and differences between R and Python (from Drew Conway):

 computingskillsforbiologists.com/learnrfrompython.

Many websites offer courses on R for free or for a fee. Typically, the first few classes are free. For example,

tryr.codeschool.com,

datacamp.com.

Find ideas and answers on R community blogs, forums, and Q&A sites:

r-bloggers.com,

stackoverflow.com,

stats.stackexchange.com.

CHAPTER 9

• • • • • • • • • • • • •

Data Wrangling and Visualization

9.1 Efficient Data Analysis and Visualization

For your research, you will need to manipulate and visualize large data sets. In previous chapters, we introduced Python and R, and showed how the analysis of data can be automated by writing simple programs. In this chapter, we cover a bundle of R packages, called the `tidyverse`, which can be used to write fast, well-organized, and very readable code for data *wrangling*,[1] analysis, and visualization. These packages introduce many features that are typical of databases (which we will cover in the next chapter), as well as a powerful framework for plotting (`ggplot2`), which produces publication-ready figures. Many of these packages were conceived and developed by Hadley Wickham— one of the leading R developers, who has contributed immensely to the growth of R and the R community.

9.2 Welcome to the `tidyverse`

The `tidyverse` is a bundle of many interrelated packages. They share the same philosophy, as well as a common data structure. Together they form an incredibly powerful and well-integrated framework for data analysis and visualization in R. In our laboratory, all data analysis and visualization is carried out using these packages.

Importantly, learning how to use one of the packages in the `tidyverse` makes it simpler to learn the next one, as all the components share the same concepts and the same language.

---

1. Data wrangling (or munging) refers to processing unstructured data with the goal of bringing it into a more structured or manageable form. Such preprocessing makes the subsequent analysis much easier.

To install the core packages of the tidyverse, simply open your R session and type

```
> install.packages("tidyverse")
```

or install the tidyverse from the "Packages" tab in RStudio. To load the core libraries, type

```
> library(tidyverse)
```

To update all the packages in the bundle, type

```
> tidyverse_update()
```

You can list all packages belonging to the tidyverse by typing

```
> tidyverse_packages()
 [1] "broom"      "dplyr"      "forcats"
 [4] "ggplot2"    "haven"      "httr"
 [7] "hms"        "jsonlite"   "lubridate"
[10] "magrittr"   "modelr"     "purrr"
[13] "readr"      "readxl"     "stringr"
[16] "tibble"     "rvest"      "tidyr"
[19] "xml2"       "tidyverse"
```

As always, we introduce the basic components of the tidyverse, and showcase its impressive capabilities, allowing you to perform complex data manipulations using very readable and concise commands. Remember to consult the documentation to explore the full capabilities of each function. At the end of the chapter, we provide resources meant to allow you to master the powerful tools of the tidyverse.

9.2.1 Reading Data

Using base R in the previous chapter, we read our data using the read.csv or read.table commands. These routines are suitable if the quantity of data is small, but can take a very long time when there is a lot. The package readr (which is part of the tidyverse) provides improved routines for reading files. Besides being much faster, these functions have better strategies to guess the type of data contained in each column of your character-separated file, they deal with dates in a more consistent manner, and they do not try to convert all your strings into factors (no more typing stringsAsFactors = FALSE)! When the data file is massive, these functions show a progress bar, so that you can see that the program is running.

The readr package provides several functions to read specific formats of text files:

read_csv   Comma-separated files
read_csv2   Semicolon-separated files
read_tsv   Tab-separated files
read_fwf   Fixed-width columns
read_delim   Any delimiter you specify (e.g., set delim = "$" for columns separated by dollar signs)

To explore the various packages in the tidyverse, we are going to use data from Fauchald et al. (2017). They tracked the population size of various herds of caribou in North America over time, and correlated population cycling with the amount of vegetation and sea-ice cover. The directory data_wrangling/data/FauchaldEtAl2017 contains the data.

To start our exploration, we launch RStudio and load the tab-separated file popsize.csv in CSB/data_wrangling/sandbox:

```
> library(tidyverse)
> setwd("CSB/data_wrangling/sandbox/")
> popsize <- read_tsv("../data/FauchaldEtAl2017/
    ↳ pop_size.csv")
```

9.2.2 Tibbles

The packages in the tidyverse are based on a new data structure called a "tibble." This is essentially a new and improved version of the data.frame object. Practically, tibbles and data frames work in the same way. However, tibbles are superior in that when printed, they display only what fits on the screen, and report basic information about the data. For example, type

```
> popsize
# a tibble: 114 x 3
     Herd  Year  Pop_Size
    <chr> <int>    <int>
1    WAH  1970    242000
2    WAH  1976     75000
3    WAH  1978    107000
4    WAH  1980    138000
5    WAH  1982    217863
6    WAH  1986    229000
7    WAH  1988    343000
8    WAH  1990    417000
9    WAH  1993    478822
10   WAH  1996    463000
# ... with 104 more rows
```

showing the header and type of the data stored in each column, as well as the
size of the tibble (114 x 3). Instead of printing all of the data, only a few lines
are printed. You can use head or tail to extract the first few or last few rows
in the tibble:

```
> head(popsize, 3)
# a tibble: 3 x 3
     Herd  Year  Pop_Size
    <chr> <int>    <int>
1    WAH  1970    242000
2    WAH  1976     75000
3    WAH  1978    107000

> tail(popsize, 3)
# a tibble: 3 x 3
     Herd  Year  Pop_Size
    <chr> <int>    <int>
1    BEV  1988    189561
2    BEV  1994    276000
3    BEV  2011    124189
```

Similar to their Unix counterparts (see section 1.5.2), the head and tail functions return a smaller tibble containing the first and last few rows, respectively, of the original tibble. Note that the row numbers refer to the tibble produced by the command, not the complete data set.

RStudio provides a neat feature to inspect the full data set. You can open the data in a spreadsheet-like environment by typing

```
> View(popsize)
```

To print a brief summary of the structure of your data, use

```
> glimpse(popsize)
Observations: 114
Variables: 3
$ Herd     <chr> "WAH", "WAH", "WAH", "WAH", "WAH", "...
$ Year     <int> 1970, 1976, 1978, 1980, 1982, 1986, ...
$ Pop_Size <int> 242000, 75000, 107000, 138000, 21786...
```

This chapter is dedicated to the tidyverse but remember that everything seamlessly integrates with base R. Hence, to extract the number of rows and columns in a tibble, you can also type

```
> dim(popsize)
[1] 114   3
> nrow(popsize)
[1] 114
> ncol(popsize)
[1] 3
```

9.3 Selecting and Manipulating Data

Using base R, subsetting the data typically requires typing many dollar signs, square brackets, and parentheses. This makes the code less readable and more error prone—especially when many brackets are nested.

In this respect, the tidyverse is a real game changer! Its dplyr package contains many functions to manipulate your data—the tidy way. Conveniently, the names of the functions already provide a clue to what they accomplish. Consult the box for an overview:

select	Select columns by name.
slice	Select rows by position.
filter	Select rows matching certain condition(s).
arrange	Sort or rearrange data.
mutate	Add variables based on existing data.
group_by	Group data according to a variable.
summarise[a]	Compute statistics on the data.

a. Hadley Wickham is originally from New Zealand, hence the British spelling of summarise. You can use the US summarize in your code with the same result.

To explore each function in more detail, we load a second file of the Fauchald et al. (2017) data set. The file ndvi.csv contains the normalized difference vegetation index (NDVI, a common measure to estimate vegetation in a certain area through remote sensing) measured at different times of the year for the home range of each herd:

```
> ndvi <- read_tsv("../data/FauchaldEtAl2017/ndvi.csv")
> head(ndvi)
# a tibble: 6 x 4
   Herd  Year NDVI_May NDVI_June_August
   <chr> <int>    <dbl>            <dbl>
1   BAT  1982  0.21440        0.3722679
2   BAT  1983  0.20448       -0.9977483
3   BAT  1984  0.24650        1.5864094
4   BAT  1985  0.24444        0.6420830
5   BAT  1986  0.20046       -0.3630283
6   BAT  1987  0.22399        0.7463858
```

9.3.1 Subsetting Data

If you want to extract only specific columns by name (e.g., Herd and NDVI_May), you can use select:

```
> select(ndvi, Herd, NDVI_May)
# a tibble: 360 x 2
   Herd NDVI_May
```

```
     <chr>     <dbl>
1     BAT   0.21440
2     BAT   0.20448
3     BAT   0.24650
4     BAT   0.24444
5     BAT   0.20046
6     BAT   0.22399
7     BAT   0.23335
8     BAT   0.23381
9     BAT   0.21823
10    BAT   0.28747
# ... with 350 more rows
```

where you first specify the tibble you want to use (e.g., `ndvi`), and then the columns you want to extract, separated by commas.

The `select` command has some useful variations, for example,

```
# use a colon to include all columns in between
> select(ndvi, Herd:NDVI_May)
# use a minus sign to exclude specific columns
> select(ndvi, -Herd, -Year)
# you can combine the two features
> select(ndvi, -(Year:NDVI_May))
# use a regular expression for the name of the column
> select(ndvi, matches("NDV"))
```

Having seen how to select certain columns, we turn to rows. The function `filter` can be used to select rows whose content matches certain conditions:

```
# select rows with value "WAH" for variable Herd
> filter(popsize, Herd == "WAH")
# select rows for years 1970 to 1980
> filter(popsize, Year >= 1970, Year <= 1980)
# select specific, nonconsecutive years
> filter(popsize, Year %in% c(1970, 1980, 1990))
```

There are additional functions to subset data by specific criteria or at random:

```
# select rows according to row number
# e.g., everything between rows 20 and 30
> slice(popsize, 20:30)
# top 10 rows when ordered by Pop_Size
> top_n(popsize, 10, Pop_Size)
# take the 15 rows with smallest Pop_Size
> top_n(popsize, 15, desc(Pop_Size))
# extract 5 rows at random
> sample_n(popsize, 5)
# 2% of the rows at random
> sample_frac(popsize, 0.02)
```

9.3.2 Pipelines

When manipulating data in base R, we tend to combine several steps by nesting function calls, to avoid assigning intermediate steps to separate variables:

```
# an example of nested base R code
unique(popsize[order(popsize$Herd), 1])
```

Understanding these commands when reading code is far from trivial (for your reference, here we have extracted all the unique herd names in the tibble popsize, and ordered them alphabetically).

Using the aptly named functions of the dplyr package improves the situation, but we are still dealing with nested calls:

```
# first, take only column Herd
> select(popsize, Herd)
# second, remove repeated values, using distinct
> distinct(select(popsize, Herd))
# finally, sort the data using arrange
> arrange(distinct(select(popsize, Herd)), Herd)
```

The third line yields the desired result. However, to make the code easier to understand and manipulate, the tidyverse provides a "pipeline" operator, similar to the vertical pipe in Unix (in section 1.6.1). Let's select and sort the herd names again:

```
> popsize %>% select(Herd) %>% distinct() %>% arrange
  ↳ (Herd)
```

The pipeline operator (%>%) takes whatever is on the left, and uses it as input for the function on the right. To type the operator in RStudio, press Ctrl+Shift+M.

Each pipeline returns a tibble, which can be used exactly like any other R object (e.g., assigned to a new variable):

```
> herds <- popsize %>%
              select(Herd) %>%
              distinct() %>%
              arrange(Herd)
```

As shown above, we can split our pipelines over several lines. Such formatting improves readability even further, making it very easy to add a piece of code in between operations, or to modify the columns you want to operate on.

Intermezzo 9.1
  (a) Extract the unique Years in popsize and order them.
  (b) Find the row in ndvi with the largest NDVI_May.
  (c) List the three years with the largest Pop_Size for the Herd called WAH. Perform this operation in two different ways.

9.3.3 Renaming Columns

To rename a column, simply call the rename function and assign the old name to a new one:

```
> popsize %>% rename(h = Herd)
```

Note, that we could have written

```
rename(popsize, h = Herd)
```

but for consistency, we will "unnest" all function calls and continue using the pipeline notation.

9.3.4 Adding Variables

To add a new variable whose content is a function of some other columns, use mutate.[2] For example, suppose you want to add a new column to the ndvi tibble, called meanNDVI, containing the average of the values in May and those in June/August:

```
> ndvi %>% mutate(meanNDVI = (NDVI_May + NDVI_June_August)
    ↳ / 2) %>% head(4)
# a tibble: 4 x 5
   Herd  Year NDVI_May NDVI_June_August   meanNDVI
  <chr> <int>    <dbl>            <dbl>      <dbl>
1   BAT  1982  0.21440        0.3722679  0.2933339
2   BAT  1983  0.20448       -0.9977483 -0.3966341
3   BAT  1984  0.24650        1.5864094  0.9164547
4   BAT  1985  0.24444        0.6420830  0.4432615
```

If you use transmute instead of mutate, you retain only the column(s) containing the results:

```
> ndvi %>% transmute(meanNDVI = (NDVI_May +
    ↳ NDVI_June_August) / 2) %>% head(4)
# a tibble: 4 x 1
   meanNDVI
      <dbl>
```

---

2. If you want to add a new variable that is independent of all other variables in your data set, use add_column. This function is part of the tibble package that is automatically loaded with the tidyverse library.

```
1  0.2933339
2 -0.3966341
3  0.9164547
4  0.4432615
```

9.4 Counting and Computing Statistics

9.4.1 Summarize Data

You can use summarise to create summaries of the data. For example, compute the average NDVI_May for the whole data set:

```
> ndvi %>% summarise(mean_May = mean(NDVI_May))
# a tibble: 1 x 1
  mean_May
     <dbl>
1 0.2522392
```

You can compute many statistics at once:

```
> ndvi %>% summarise(mean_May = mean(NDVI_May),
                     sd_May = sd(NDVI_May),
                     median_May = median(NDVI_May))
# a tibble: 1 x 3
  mean_May      sd_May median_May
     <dbl>       <dbl>      <dbl>
1 0.2522392 0.09046798  0.2526943
```

Within summarise, you can perform any operation, including user-defined functions.

9.4.2 Grouping Data

The power of the dplyr package, however, is that you can perform operations like mutate and summarise on *grouped* data. You use the function group_by to define a grouping for your data, and then mutate or summarise will be applied to each group separately. For example, let's compute the average, minimum, and maximum population sizes for each herd, taken across all years:

```
> popsize %>% group_by(Herd) %>%
              summarise(avgPS = mean(Pop_Size),
              minPS = min(Pop_Size),
              maxPS = max(Pop_Size)) %>%
              arrange(Herd)
# a tibble: 11 x 4
     Herd      avgPS  minPS   maxPS
     <chr>     <dbl>  <int>   <int>
1    BAT  212715.833  31900  486000
2    BEV  179472.375 124000  276000
3    BLW   57650.500  17897  112360
4    CAH   27504.818   5000   70034
5    CBH    8121.875   1821   19278
6    GRH  289731.000  14200  775891
7    LRH  284571.429  56000  628000
8    PCH  141853.846  99959  197000
9    QAM  209851.222  40000  496000
10   TCH   29155.545   3500   64106
11   WAH  308895.688  75000  490000
```

We have performed the summaries for each herd separately. To count the number of rows belonging to each group, use `tally`:

```
> popsize %>% group_by(Herd) %>% tally() %>% arrange(Herd)
# a tibble: 11 x 2
     Herd      n
     <chr> <int>
1    BAT     12
2    BEV      8
3    BLW      8
4    CAH     11
5    CBH      8
6    GRH     11
7    LRH      7
8    PCH     13
```

```
 9     QAM      9
10     TCH     11
11     WAH     16
```

dplyr also provides the function n, which counts rows. As such, the same operation can be carried out using

```
> popsize %>%
      group_by(Herd) %>%
      summarise(tot = n()) %>%
      arrange(Herd)
```

To show how you can use mutate (or transmute) on grouped data, we are going to compute a *z*-score for the population size of each herd in each year. For each herd and year, we compute the difference between the population size in that year and the mean population size, and divide the difference for the corresponding standard deviation:

```
> popsize %>%
      group_by(Herd) %>%
      mutate(zscore = (Pop_Size - mean(Pop_Size)) / sd(
          ↳ Pop_Size))
Source: local data frame [114 x 4]
Groups: Herd [11]

    Herd  Year Pop_Size      zscore
   <chr> <int>   <int>       <dbl>
1    WAH  1970   242000 -0.4951977
2    WAH  1976    75000 -1.7314212
3    WAH  1978   107000 -1.4945401
4    WAH  1980   138000 -1.2650614
5    WAH  1982   217863 -0.6738727
6    WAH  1986   229000 -0.5914307
7    WAH  1988   343000  0.2524584
8    WAH  1990   417000  0.8002460
9    WAH  1993   478822  1.2578856
10   WAH  1996   463000  1.1407627
# ... with 104 more rows
```

You can accomplish the same operation using the function scale:

```
> popsize %>%
        group_by(Herd) %>%
        mutate(zscore = scale(Pop_Size))
```

Intermezzo 9.2

(a) Compute the average Pop_Size for each Herd in popsize.

(b) Identify the Herd with the largest standard deviation for NDVI_May.

(c) Add a new column containing the population size relative to the mean population size (i.e., Relative_Pop should be 1 when the population in that year is exactly as large as the mean population, a value of 2 if the population is twice the mean, 0.5 if it's half of the mean, etc.).

9.5 Data Wrangling

All the packages in the tidyverse are based on the notion of "tidy" data, meaning that

(a) each variable is stored its own column;

(b) each set of observations is stored in its own row;

(c) each table contains a consistent set of data (e.g., all data on sampling sites in one table, all data on climate in another);

(d) multiple tables can be joined if they contain a common column (e.g., the site-description table has a column site and the table on climate also has a column site so that the two tables can be linked).

Basically, the tidy format is the result of an operation called "normalization," which we will explore in depth in the next chapter, where we work with databases.

Because you may need to work with data organized in very different ways, the tidyverse package tidyr is entirely dedicated to converting them into a tidy form. In this section, we present some of the most common functions of the tidyr package. If you loaded the complete tidyverse above, you are ready to wrangle some data; otherwise you can load the tidyr library individually.

9.5.1 Gathering

The first example we're going to examine is one in which the data have been compiled in a tabular form that is well suited to being published in a paper, but difficult to handle for data analysis. Load the file sea_ice.csv of the

Fauchald et al. (2017) data set. For each Year and Herd, the proportion of surface covered by ice is reported for each month:

```
> seaice <- read_tsv("../data/FauchaldEtAl2017/sea_ice.csv
  ↳ ")
> head(seaice)
# a tibble: 6 x 14
  Herd  Year   Jan   Feb   Mar   Apr   May   Jun   Jul
  <chr> <int> <dbl> <dbl> <dbl> <dbl> <dbl> <dbl> <dbl>
1 WAH   1979  91.7  98.0  97.1  93.0  84.4  74.9  64.4
2 WAH   1980  98.6  98.4  97.5  95.9  88.5  76.7  58.1
3 WAH   1981  98.3  98.1  97.2  95.3  87.7  72.5  60.0
4 WAH   1982  98.3  97.9  97.9  97.8  92.2  73.9  57.7
5 WAH   1983  98.4  98.6  98.5  96.8  87.8  80.7  71.9
6 WAH   1984  95.7  97.9  97.4  97.4  91.6  76.2  64.5
# ... with 5 more variables: Aug <dbl>, Sep <dbl>,
#   Oct <dbl>, Nov <dbl>, Dec <dbl>
```

This is a very compact form to present the data in, but not very suitable for computing. Note how the rule "each set of observations is stored in its own row" is violated. We would like to organize the data in a tidy tibble with four columns: Herd, Year, Month, and Cover. To this end, we gather columns 3 to 14 in the tibble, and use the current column name (i.e., name of month) as a value for the new variable Month. The current contents of each cell will be provided as a value to the new variable Cover:

```
> seaice %>% gather(Month, Cover, 3:14)
# a tibble: 4752 x 4
  Herd  Year Month Cover
  <chr> <int> <chr> <dbl>
1 WAH   1979  Jan   91.7
2 WAH   1980  Jan   98.6
3 WAH   1981  Jan   98.3
4 WAH   1982  Jan   98.3
5 WAH   1983  Jan   98.4
6 WAH   1984  Jan   95.7
7 WAH   1985  Jan   95.0
8 WAH   1986  Jan   97.0
```

```
9    WAH  1987   Jan  98.2
10   WAH  1988   Jan  98.6
# ... with 4742 more rows
```

We now see our data organized in tidy form, with each observation in its own row. We can overwrite the previous tibble:

```
> seaice <- seaice %>% gather(Month, Cover, 3:14)
```

Intermezzo 9.3
   (a) Compute the average sea-ice cover for each month and herd, by averaging over the years.
   (b) Read the data file again and try to repeat the analysis with the data in the original format (i.e., before calling gather). This will illustrate the value of organizing data in a tidy format.

9.5.2 Spreading

As stated above, the tidy form is exceptionally good for computing, but not so good for human consumption. For this reason, tidyr provides a function that is the "opposite" of gather, called spread. For example, let's produce a table in which for each Herd (row) and Year (column), we report the population size.

Our tibble is currently organized as

```
> head(popsize, 3)
# a tibble: 3 x 3
   Herd  Year Pop_Size
  <chr> <int>    <int>
1   WAH  1970   242000
2   WAH  1976    75000
3   WAH  1978   107000
```

We want to turn the Year(s) into columns, and use the Pop_Size values to fill the cells. To keep the table small, let's consider only the years between 1980 and 1984:

```
> popsize %>%
        filter(Year > 1979, Year < 1985) %>%
        spread(Year, Pop_Size)
# a tibble: 9 x 6
   Herd '1980' '1981' '1982' '1983' '1984'
 * <chr>  <int>  <int>  <int>  <int>  <int>
 1   BAT 140000     NA 174000     NA 384000
 2   BEV 130000     NA 164338     NA 263691
 3   CAH     NA   8537     NA  12905     NA
 4   GRH 390100     NA 360450     NA 586600
 5   LRH     NA     NA     NA 101000     NA
 6   PCH     NA     NA 125174 135284     NA
 7   QAM  40000     NA 180000 234000     NA
 8   TCH     NA     NA     NA     NA  11822
 9   WAH 138000     NA 217863     NA     NA
```

Note that spread fills the cells for which we have no information with NA by default. You can choose a custom fill value by setting fill within the spread call:

```
> popsize %>%
        filter(Year > 1979, Year < 1985) %>%
        spread(Year, Pop_Size, fill = "")
```

9.5.3 Joining Tibbles

When working with multiple tables belonging to the same project, you might need to combine them—an operation called join. This concept is borrowed from databases, which we will explore in more detail in the next chapter. Briefly, an inner_join is an operation in which we take two tables containing common columns, and produce a third table linking the rows of the first table with those of the second according to their shared columns (see figure 10.3). This is called an *inner join* because a row in the new table is produced whenever it has a corresponding row in both the first *and* second tables. Thus, the inner join works like an AND function. An *outer join*, on the other hand, produces a row in the third table whenever there is a corresponding row in the first *or* second table. There are multiple variants of an outer join, described in detail in the documentation of dplyr::join.[3]

For example, we want to correlate the NDVI_May value with the Pop_Size for each year. However, these values are contained in two different tibbles. We therefore join popsize and ndvi using their common columns Herd and Year:

```
> combined <- inner_join(popsize, ndvi)
Joining, by = c("Herd", "Year")
> head(combined, 4)
# a tibble: 4 x 5
   Herd  Year Pop_Size  NDVI_May NDVI_June_August
   <chr> <int>   <int>     <dbl>            <dbl>
1   WAH  1982  217863 0.2894798        1.3849703
2   WAH  1986  229000 0.3014125       -0.8109645
3   WAH  1988  343000 0.3758451        1.0407032
4   WAH  1990  417000 0.4586601        4.5864858
> dim(popsize)
[1] 114    3
> dim(ndvi)
[1] 360    4
> dim(combined)
[1] 81   5
```

Note that our combined data set has fewer rows than either of the two tibbles we joined—inner_join creates a row only when there is a corresponding row in both tables.

Now that we have a combined data set, we can compute the correlation between Pop_Size and NDVI_May using the function cor:

```
> cor(combined %>% select(Pop_Size, NDVI_May))
            Pop_Size    NDVI_May
Pop_Size   1.0000000  -0.2094812
NDVI_May  -0.2094812   1.0000000
```

---

3. See computingskillsforbiologists.com/joiningtibbles.

Sometimes, you need to join two tables in which the common column has different names in the two tables. While it is best to be consistent (and simply rename the column), you can manually specify the names of the common columns by adding `by = c("col_name1" = "col_name2")` to your `inner_join` function call.

Intermezzo 9.4

(a) Produce a tibble (`avg_Perc_seaicecover`) containing the average population size and the average sea-ice cover for each combination of `Herd` and `Year`.

(b) Take a look at the documentation of `dplyr::join` and identify the command that provides a tibble with the 17 rows that are present in `popsize` but not in `Perc_seaicecover` (i.e., that do not have data for sea-ice cover).

(c) For each `Herd`, compute Kendall's rank correlation between the percentage of sea-ice cover in March and the week in which the ground snow melted (contained in `snow.csv`) for each of the years.

9.6 Data Visualization

The most salient feature of scientific graphs should be clarity. Each figure should make crystal clear (a) what is being plotted; (b) what the axes are; (c) what the colors, shapes, and sizes represent; (d) the message you want to convey. Each figure is accompanied by a (sometimes long) caption, where the details can be explained further, but the main message should be clear from glancing at the figure alone. Often, figures are the first thing editors and referees look at in a scientific article.

Many scientific publications contain very poor graphics: labels are missing, scales are unintelligible, there is no explanation of some graphical elements. Moreover, some color graphs are impossible to understand if printed in black and white, or difficult to discern for color-blind people.

Given the effort that you put into your science, you want to ensure that it is well presented and accessible. The time investment in mastering plotting software will be rewarded by clean, beautiful graphics that convey a clear message. Here we show how to draw publication-quality figures in R using the package `ggplot2`, which is also part of the `tidyverse` and therefore integrates very well with the other components illustrated above.

You might notice that, quite paradoxically, this chapter on visualization does not contain a single figure. The irony of it amuses us, but we decided that we prefer to offer this book at a reasonable price, rather than producing a more expensive volume displaying all graphs in glossy detail. By executing the commands, you will see the graphs on your screen. Furthermore, we provide all code and resulting graphics in the directory CSB/data_wrangling/all_graphs. You can choose to follow along using the file all_graphs in the formats R Markdown (.Rmd), plain R code (.R), or PDF (.pdf).

9.6.1 Philosophy of ggplot2

Unlike many other plotting systems, ggplot2 is deeply rooted in a "philosophical" vision. The goal is to conceive a grammar for all graphical representation of data.

Leland Wilkinson and collaborators (Wilkinson, 2006) proposed the *grammar of graphics*. It follows the idea of a well-formed sentence that is composed of a subject, a predicate, and an object. Likewise, the *grammar of graphics* aims to describe a well-formed graph using a grammar that captures a very wide range of statistical and scientific plots. This might be more clear with an example—take a simple two-dimensional scatter plot: How can we describe it? We have

data: This is simply data we want to plot.

mapping: What part of the data is associated with a particular visual feature? For example, which column is associated with the x-axis? Which with the y-axis? Which column corresponds to the shape or the color of the points? In ggplot2 lingo, these are called "aesthetic mappings" (aes).

geometry: Do we want to draw points? Lines? In ggplot2 we speak of "geometries" (geom).

scale: Do we want the sizes and shapes of the points to scale according to some value? Linearly? Logarithmically? Which palette of colors do we want to use?

coordinates: We need to choose a coordinate system (e.g., Cartesian, polar).

faceting: Do we want to produce different panels, partitioning the data according to one of the variables?

This basic grammar can be extended by adding statistical transformations of the data (e.g., regression, smoothing), multiple layers, adjustment of position (e.g., stack bars instead of plotting them side by side), annotations, and so on.

Exactly like in the grammar of a natural language, we can easily change the meaning of a "sentence" by adding or removing parts. Also, it is very easy to completely change the type of geometry if we are moving from, say, a histogram to a box plot or a violin plot, as these types of plots are meant to describe one-dimensional distributions. Similarly, we can go from points to lines, by changing one "word" in our code. Finally, the look and feel of the graphs is controlled by a theme system, separating the content from the presentation.

9.6.2 The Structure of a Plot

Here we focus on the function ggplot, which allows you to control all aspects of the plotting. The package also provides a simplified version of this function, called *quick plot* (qplot), which is described in the documentation.

As mentioned above, we are providing only the commands that you need to call to produce the graphs, but not the resulting graphics. However, all commands and resulting graphs are provided in the file all_graphs.Rmd in the directory CSB/data_wrangling/all_graphs.

We continue working with the data published by Fauchald et al. (2017). We now also load the file snow.csv:

```
# load the library
library(tidyverse)
# read the data
popsize <- read_tsv("../data/FauchaldEtAl2017/pop_size.csv
    ↳ ")
ndvi <- read_tsv("../data/FauchaldEtAl2017/ndvi.csv")
seaice <- read_tsv("../data/FauchaldEtAl2017/sea_ice.csv")
snow <- read_tsv("../data/FauchaldEtAl2017/snow.csv")
# convert to tidy form
seaice <- seaice %>% gather(Month, Cover, 3:14)
```

Let's build our first graph by adding the elements described in the the grammar of graphics, one by one. Note that we can expect a meaningful plot only when all elements have been declared, and we have built a well-formed "sentence."

First we need the data. Typing

```
> ggplot(data = popsize)
```

produces an empty graph, as we have set the data but have not specified what the aesthetic mappings should be. We can associate the *x*-axis with the Year, the *y*-axis with Pop_Size, and the color of the points with Herd by typing

```
> ggplot(data = popsize) +
        aes(x = Year, y = Pop_Size, colour = Herd)
```

Notice that we have added the aes using the + sign, which has been repurposed to collate the various plotting commands. Still we don't have a full graph, because we need to specify a "geometry." For example, let's draw points connected by segments:

```
> ggplot(data = popsize) +
        aes(x = Year, y = Pop_Size, colour = Herd) +
        geom_point() +
        geom_line()
```

That's it. We have created a well-formed "sentence" (data + mappings + geometry), and as a result we have produced our first fully fledged plot.

When choosing a geom, it is important to think first about the type of data you want to plot. How many variables? Are these continuous or discrete variables? *Continuous* refers to variables that can in principle take infinitely many ordered values: for example, weights, lengths, or time. Such variables are stored as numbers or date–time; ggplot2 will automatically treat these data classes as continuous variables. *Discrete* variables, on the other hand, are used for nominal, categorical, or ordinal data: the name of a treatment, of a species, and of a month are all examples of data that can be represented as discrete variables.

9.6.3 Plotting Frequency Distribution of One Continuous Variable

Let's start by exploring the geometries one can use to represent a single, continuous variable. For example, let's plot the histogram of NDVI_May across all years and sites:

```
> ggplot(data = ndvi) +
        aes(x = NDVI_May) +
        geom_histogram()
```

where each bar represents the number of data points (count) that fall into a certain data range (bin).

Note that one can define the aes of a plot in several ways. One can set it separately, within the call to ggplot, or within the call to geom:

```
# nested or "unrolled"
# these three commands produce the same graph:
> ggplot(data = ndvi) +
        aes(x = NDVI_May) +
        geom_histogram()
> ggplot(data = ndvi, aes(x = NDVI_May)) +
        geom_histogram()
> ggplot(data = ndvi) +
        geom_histogram(aes(x = NDVI_May))
```

You can interpolate the histogram and produce a density plot:

```
> ggplot(data = ndvi) +
        aes(x = NDVI_May) +
        geom_density()
```

showing that it is easy to move between different geometries if they are meant to represent the same type of data.

9.6.4 Box Plots and Violin Plots

Box plots and violin plots are excellent for providing information on the distribution of your data. A box plot marks the median (50th percentile) of the values as a line in the middle of a box enclosing half of your data (from the 25th to the 75th percentile). The lines extending from the box (whiskers) represent 1.5 × the length of the box (or interquartile range, IQR). Data points falling beyond the whiskers are drawn separately as outliers.

Violin plots add information on the distribution of data (i.e., where a violin plot is wider, more data points cluster—you can think of a violin plot as the union of two mirrored density plots).

Let's plot the distribution of NDVI_May for each herd across the years. For a box plot, we specify a discrete *x*-axis (the Herd, in this case), and a continuous *y*-axis (in this case, NDVI_May). Importantly, we can store the common part of a ggplot graph into a variable and add additional elements as needed:

```
> pl <- ggplot(data = ndvi) + aes(x = Herd, y = NDVI_May)
> pl + geom_boxplot()
> pl + geom_violin()
```

To change the color for the boxes, set the aesthetic fill:

```
> pl + geom_boxplot() + aes(fill = Herd)
```

9.6.5 Bar Plots

ggplot2 provides different geometries for bar plots. The geom_bar is used to represent count data of discrete variables. For example, we can make sure that our data for population WAH is complete by plotting the number of monthly measures contained in the data set for each year:

```
> ggplot(data = seaice %>% filter(Herd == "WAH")) +
        aes(x = Year) +
        geom_bar()
```

showing that we have 12 records per year for all the years between 1979 and 2014.

When we want the height of the bars to represent a value in a column (instead of a count), we can use geom_col:

```
> ggplot(data = seaice %>%
        filter(Herd == "WAH", Year == 1990)) +
        aes(x = Month, y = Cover) +
        geom_col()
```

As you can see, we are not quite finished: the month labels are considered strings, and therefore ordered alphabetically. We can turn them into factors, and order them in the right way:

```
> seaice$Month <- factor(seaice$Month, month.abb)
> ggplot(data = seaice %>%
        filter(Herd == "WAH", Year == 1990)) +
      aes(x = Month, y = Cover) +
      geom_col()
```

Now the plot makes much more sense!

9.6.6 Scatter Plots

Scatter plots show the relationship between two continuous variables. Let's plot the population dynamics of the herd WAH in time:

```
> pl <- ggplot(data = popsize %>%
              filter(Herd == "WAH")) +
            aes(x = Year, y = Pop_Size) +
            geom_point()
> show(pl)
```

The function show is used to display the plot stored in pl. We might want to add a smoothing function:

```
> pl + geom_smooth()
```

By default, ggplot2 uses a local regression (LOESS). We can, however, choose a different way to compute the fitting curve by specifying a method and optional parameters. For example,

```
# use a linear model
> pl + geom_smooth(method = "lm")
# use a polynomial regression
```

```
> pl + geom_smooth(method = "lm",
        formula = y ~ poly(x, 3), se = FALSE)
```

By setting se = FALSE, we suppress plotting the standard error.

9.6.7 Plotting Experimental Errors

Providing appropriate information on experimental errors is a hallmark of any credible scientific graph. Choose a type of error based on the conclusion that you want the reader to draw. While the standard deviation (SD) represents the dispersion of the data,[4] the standard error of the mean (SEM) and confidence intervals (CI) report the certainty of the estimate of a value (e.g., certainty in estimating the mean).[5]

Let's calculate some summary statistics and errors for the population sizes of two of the herds. We then plot the mean population size as a bar plot without error bars:

```
# calculate summary stats and errors for herds GRH and PCH
stats <- popsize %>% filter(Herd %in% c("GRH", "PCH")) %>%
        group_by(Herd) %>%
        summarise(
            meanPopSize= mean(Pop_Size),
            SD = sd(Pop_Size),
            N = n(),
            SEM = SD/sqrt(N),
            CI = SEM * qt(0.975, N-1))
# plot mean population size as bar plot
ggplot(data = stats) +
        aes(x = Herd, y = meanPopSize) +
        geom_col()
```

From this plot, someone might easily conclude that the mean population size of these two herds is different. Let's set up the error bars and plot again:

---

4. If you consider the distribution of your data to be the most important point, a scatter plot, box plot, or violin plot is more appropriate than a bar plot.

5. Cumming et al. (2007) provide a brief explanation of error bars in experimental biology.

```
# set up aesthetic mapping for confidence intervals
limits <- aes(ymax = stats$meanPopSize + stats$CI,
              ymin = stats$meanPopSize - stats$CI)
# plot including confidence intervals
ggplot(data = stats) +
       aes(x = Herd, y = meanPopSize) +
       geom_col() +
       geom_errorbar(limits, width = .3)
```

The graph including the 95% confidence intervals supports the idea that the mean population sizes of these two herds are not significantly different (a *t*-test would confirm that much). To prevent misinterpretation, always state which type of error is shown in your graph.

For more information on plotting means and errors in R, have a look at the *Cookbook for R* by Winston Chang.[6] This book also provides a helpful function to automate your error calculations.

Intermezzo 9.5

(a) Explore the data set snow and plot the Perc_snowcover versus Week_snowmelt. Choose an appropriate geom.

(b) For each year, compute the average Perc_snowcover and Week_snowmelt and create a scatter plot.

(c) Produce a box plot of the Pop_Size for each Herd.

(d) Create a scatter plot showing the average population size (taken across herds) including standard deviation, for the years 2008 through 2014.

9.6.8 Scales

Scales are used to modify the *x*- and *y*-axes, as well as to determine how colors, sizes, and shapes behave. You can set a scale for each aesthetic mapping. We have already seen that you can use aes to associate certain columns in your tibble with the *x*- and *y*-coordinates. Here are a few other aesthetic mappings that you can modify using the corresponding scales:

---

6. computingskillsforbiologists.com/plottingerrors.

fill   Color filling an object (e.g., bars of a histogram, boxes of a box plot)
colour   Color of the border of an object (e.g., boxes of a box plot) or color of
         a point or line
size   Size of a point
shape   Shape of a point
alpha   Level of transparency of a point or a bar

One important feature of scales is that we need to distinguish between continuous and discrete variables. Let's start with a discrete scale that controls the fill. When you draw a box plot of the population size of each herd,

```
> pl <- ggplot(data = popsize,
        aes(x = Herd, y = Pop_Size, fill = Herd)) +
        geom_boxplot()
> show(pl)
```

ggplot2 uses the default colors. You can alter these colors by changing the scale for the aesthetic mapping fill.

```
# choose a palette from ColorBrewer
> pl + scale_fill_brewer(palette = "Set3")
# palette based on hue
> pl + scale_fill_hue()
# manually set values and rename the legend
> pl + scale_fill_manual(values = rainbow(11),
        name = "aaa")
```

Similarly, you can set other scales such as size or colour. Take the graph

```
> pl <- ggplot(data = seaice %>% filter(Herd == "BEV")) +
        aes(x = Year, y = Month, colour = Cover,
        size = Cover) +
        geom_point()
> show(pl)
```

where we are setting a continuous scale for the color, as well as a discrete scale for the size of the point. You can change the colors by setting, for example,

```
> pl + scale_colour_gradient(high = "white", low = "red")
```

A word of caution when choosing colors: what looks pretty to you might be unintelligible for a color-vision-impaired reviewer (1 in 12 men and 1 in 200 women). Humans also tend to perceive colors with a different weight (i.e., importance). By choosing your colors haphazardly, you might unknowingly give more weight to unimportant details. Using the ColorBrewer palettes or similar color schemes minimizes this effect.[7]

Here, we briefly introduced scales that change the color of visual elements. There are, however, many more scales, for example, those transforming the appearance of the axes. Some examples are scale_x_log10 for showing the x-axis in a logarithmic scale or scale_x_reverse to reverse the axis. For more complex transformations, use scale_x_continuous and scale_x_discrete. Corresponding commands exist for transformation of the y-axis.

9.6.9 Faceting

Sometimes you want to split data according to one or more variables, but still plot the resulting panels in one plot for easy comparison. In ggplot2 this operation is called *faceting*. Two functions are available to facet data: facet_wrap and facet_grid. While facet_wrap arranges panels sequentially and omits empty panels when a certain combination of variables is not available, facet_grid arranges all variable combinations in a grid regardless of data availability.

For example, let's see how sea-ice dynamics have changed through the decades. We take two populations (WAH and BAT), and the years 1970, 1990, and 2010. For each combination, we produce a bar plot as in the previous example:

```
> ggplot(data = seaice %>%
                filter(Herd %in% c("WAH", "BAT"),
                Year %in% c(1980, 1990, 2000, 2010))) +
```

---

7. Find some guidance on producing better scientific graphics for color-vision-impaired people at computingskillsforbiologists.com/designguide.

```
        aes(x = Month, y = Cover) +
        geom_col() +
        facet_grid(Year~Herd)
```

In a facet_grid, the panels have common *x*- and *y*-axes, and are ordered according to up to two variables (in our example the variable Year is mapped into the rows of the grid, and Herd into the columns). Notice that sea-ice cover in September decreased dramatically over four decades.

You can use facet_wrap when you want simply to fit several panels on the page, but the ordering of the panels has no particular meaning. For example, the sea-ice cover in 2010 for all populations can be plotted using

```
> ggplot(data = seaice %>% filter(Year == 2010)) +
        aes(x = Month, y = Cover) +
        geom_col() +
        facet_wrap(~Herd)
```

One nice feature of facet_wrap is that it allows for the ranges of the *x*- and *y*-axes to change between the panels (very useful when they vary considerably):

```
> ggplot(data = seaice %>% filter(Year == 2010)) +
        aes(x = Month, y = Cover) +
        geom_col() +
        facet_wrap(~Herd, scales = "free")
```

This, of course, requires more space, but is worth considering for graphs that would be impossible to read otherwise.

9.6.10 Labels

We can change the labels of the graphs via xlab, ylab, and ggtitle:

```
> pl <- ggplot(data = popsize) +
        aes(x = Year, y = Pop_Size) +
        geom_point()
```

```
> pl + xlab("Year")
> pl + ylab("Population Size")
> pl + ggtitle("Population Dynamics")
```

9.6.11 Legends

By default, ggplot2 will place all the legends on the right of your plot. You can move them all to another position by specifying the legend.position option:

```
> pl <- ggplot(data = popsize) +
        aes(x = Herd, y = Pop_Size, fill = Herd) +
        geom_boxplot()
# default
> show(pl)
# move legend
> pl + theme(legend.position = "bottom")
> pl + theme(legend.position = "top")
# remove legend
> pl + theme(legend.position = "none")
```

ggplot2 strives to keep the number of legends as small as possible. For example, if you associate a certain column of your data with both the size and the colour of your points, ggplot2 will create only one legend for these features. If, however, different features of the data are mapped into different aesthetics, a separate legend will be created for each feature.

Importantly, each legend comes with a *guide*. You can use the command guide to change the title of a legend, alter its features, or to suppress it altogether. For example,

```
> pl <- ggplot(data = popsize) +
        aes(x = Year, y = Pop_Size, colour = Herd,
                alpha = sqrt(Pop_Size)) +
        geom_point()
```

```
> show(pl)
> pl + guides(colour = guide_legend(nrow = 4,
              title = "herd"),
        alpha = guide_legend(direction = "horizontal",
              title = "al"))
# suppress only one legend
> pl + guides(colour = "none")
```

Read the documentation for guide_legend:[8] there are many parameters that you can adjust to make your legends look just right.

9.6.12 Themes

Once we have built a plot we like, we can change its general appearance by choosing a different theme. Themes allow you to quickly and consistently adapt your plots to printing requirements, etc. Here are some examples:

```
# load data if not done for intermezzo
snow <- read_tsv("../data/FauchaldEtAl2017/snow.csv")
pl <- ggplot(data = snow %>%
        filter(Herd == "CAH"),
        aes(y = Week_snowmelt, x = Perc_snowcover)) +
        geom_point()
# default
show(pl)
# black and white (light background)
pl + theme_bw()
# line draw
pl + theme_linedraw()
# minimalist theme
pl + theme_minimal()
```

The package ggthemes provides several extra themes, including themes that mimic the style of popular publications:

```
library(ggthemes)
# Wall Street Journal
show(pl + theme_wsj())
# five thirty-eight
show(pl + theme_fivethirtyeight())
```

You can also easily create your own themes,[9] and many more are available online.

9.6.13 Setting a Feature

Sometimes we want to set one of the features of the graph, rather than mapping some data to it. For example, we want to have a certain size, color, shape, or transparency for all the points. To do so, we need to set the value outside the aesthetics:

```
# use color as an aesthetic mapping, associated with Herd
> pl <- ggplot(data = popsize) +
        aes(x = Year, y = Pop_Size, colour = Herd) +
        geom_point()
# set color to be red for all points
> pl <- ggplot(data = popsize) +
        aes(x = Year, y = Pop_Size) +
        geom_point(colour = "red")
```

Similarly, you can set the general size, shape, fill, and alpha for your visual elements.

9.6.14 Saving

Use the function ggsave to save a plot:

```
> ggsave(filename = "../sandbox/test.pdf", plot = pl,
         width = 3, height = 4)
```

---

9. See computingskillsforbiologists.com/ggplotthemes for an overview.

You can also set `plot = last_plot()` to save a plot without first assigning it to a variable. You can choose from various formats (e.g., PNG, SVG, JPEG), set the resolution for raster formats, and specify dimensions in inches, centimeters, or pixels.

Intermezzo 9.6

(a) For each herd (facet), produce a graph showing the histogram of the week in which ground snow melts.

(b) A geometry we didn't explore is `geom_tile`, which can be used to produce heat maps. Explore its documentation[a] and draw a heat map having `Year` as the x-axis, `Herd` as the y-axis, and `Week_snowmelt` as the color of each cell.

(c) Produce a scatter plot showing the `Perc_snowcover` vs. `Week_snowmelt`. Facet using `Herd`, and add a smoothing line to each panel.

---

a. computingskillsforbiologists.com/rectangles.

9.7 Tips & Tricks

- If a column name is a number, or contains white spaces, you can refer to it in the code by using backticks as quotes:

```
> popsize %>%
        filter(Year > 1979, Year < 1985) %>%
        spread(Year, Pop_Size) %>%
        select(Herd, '1980')
```

- Sometimes, you need to remove the grouping information before proceeding with your pipeline. To remove all the information on groups from a tibble, pass it to `ungroup`.

```
# to see the difference, run the code with
# and without ungroup
> popsize %>%
        group_by(Herd, Year) %>%
```

```
        tally() %>%
        ungroup() %>%
        summarise(n = sum(n))
```

- Operations by row. We have performed operations involving columns (or subsets of columns). When you want to perform an operation by row you can use rowwise. Suppose you want to generate a new column that contains the highest values of NDVI_May and NDVI_June_August, respectively, for each year and herd. The following does not work:

```
> ndvi %>%
        mutate(maxndvi = max(NDVI_May,
                NDVI_June_August)) %>%
        head(4)
# a tibble: 4 x 5
   Herd  Year NDVI_May NDVI_June_August  maxndvi
   <chr> <int>    <dbl>            <dbl>    <dbl>
1   BAT  1982  0.21440        0.3722679 6.308922
2   BAT  1983  0.20448       -0.9977483 6.308922
3   BAT  1984  0.24650        1.5864094 6.308922
4   BAT  1985  0.24444        0.6420830 6.308922
```

This is because the maximum is taken over all the values in each column. In such cases, you can apply the function row by row:

```
> ndvi %>%
        rowwise() %>%
        mutate(maxndvi = max(NDVI_May,
                NDVI_June_August)) %>%
        head(4)
# a tibble: 4 x 5
   Herd  Year NDVI_May NDVI_June_August   maxndvi
   <chr> <int>    <dbl>            <dbl>     <dbl>
1   BAT  1982  0.21440        0.3722679 0.3722679
2   BAT  1983  0.20448       -0.9977483 0.2044800
3   BAT  1984  0.24650        1.5864094 1.5864094
4   BAT  1985  0.24444        0.6420830 0.6420830
```

9.8 Exercises

9.8.1 Life History in Songbirds

Martin (2015) studied songbirds in temperate and tropical environments. He showed (figure 2A) that peak growth rate is higher in species suffering higher nest predation risk, and is lower in tropical species with the same level of risk as temperate species. In the same figure (2B) he reported that nestling period covaries with growth rate, with tropical species having shorter nestling periods (for the same growth rate) than temperate species.

The file Martin2015_figure2.pdf contains a figure generated with ggplot2 similar to figure 2 of the original paper. Reproduce the figure using the file Martin2015_data.csv deposited in the CSB/data_wrangling/data directory.

9.8.2 Drosophilidae Wings

Bolstad et al. (2015) studied the allometric relationships between wing length of Drosophilidae and the length of the L2 vein that runs across the wing. They measured more than 20,000 individuals, belonging to 111 species. In their figure 1, they show regressions between the log length of the wing size and the log length of the L2 vein. They produce a regression for each species and sex. They then added points showing the average values for each species. The file data/Bolstad2015_figure1.pdf contains a simplified version of figure 1 of the original paper. Reproduce the figure. The data are stored in CSB/data_wrangling/data/Boldstad2015_data.csv. The logarithms of wing size and L2 length are already taken.

9.8.3 Extinction Risk Meta-Analysis

Urban (2015) conducted a meta-analysis of extinction risk and its relationship to climate change. He included 131 studies. In figure 1, he plotted the number of studies reporting a certain overall proportion of extinction risk. The data (data/Urban2015_data.csv) are at a finer resolution than needed for this figure. In fact, each study has been split into different lines according to the method and taxa used to compute the extinction risk. To reproduce figure 1, you will need to summarize data by grouping lines with the same author/year, and for each study compute the proportion of species at risk of extinction (sum the N.Ext for each study, and divide by the corresponding sum of Total.N). Close inspection of the original figure shows that the data have been plotted in bins of unequal size (e.g., 0.5 < proportion < 1 is in one bin) so you will need to classify the various proportions into appropriate bins (0, 0–0.05, 0.05–0.1, ..., 0.5–1) before plotting. A ggplot2 version of figure 1 of the original paper is reported in data/Urban2015_figure1.pdf. Reproduce the figure.

9.9 References and Reading

Books

G. Grolemund & H. Wickham *R for Data Science*, O'Reilly, 2017.
  A concise and crystal-clear introduction to the tidyverse and data analysis in general. Also available for free at r4ds.had.co.nz.

Online Resources

The tidyverse website, with short descriptions of all core packages:
  tidyverse.org.

The definitive style guide for the tidyverse:
  style.tidyverse.org.

Extremely well-designed documentation for ggplot2:
  ggplot2.tidyverse.org/reference.

Several good video tutorials are available on YouTube:
  computingskillsforbiologists.com/tidyversevideo,
  computingskillsforbiologists.com/videohadley.

A nice list with many extension packages for ggplot2:
  computingskillsforbiologists.com/ggplot2ext.

CHAPTER 10

• • • • • • • • • • • • •

Relational Databases

10.1 What Is a Relational Database?

A database is a structured collection of data, making the data easy to access, manage, and update. Databases are particularly useful for handling large data sets, and are one of the main items in the toolbox of biologists working with sequence data, spatial data (GIS,[1] combining maps and databases), and "big data" in general.

This chapter introduces relational databases, in which data are arranged in tables, and different tables are connected through related columns. Relational databases consist of three components: first, the data, organized in tables; second, a language to query the data (Structured Query Language, SQL); and third, a program to manage the data (relational database management system, RDBMS).

The tables of a relational database can be thought of as a series of spreadsheets that are linked with each other. Each table is composed of rows (called *records*) and columns (called *fields*). Each field contains a certain type of data (e.g., text, numeric, date). The relationships between the tables are encoded in a *schema*.

To interact with a relational database, you write SQL *queries*, commands that retrieve, update, or modify the data. In most cases, running a query returns a particular representation of a (processed) subset of the data.

Relational database management systems store the data and related objects in specific binary formats that can be read by specialized software (as opposed to text files, which can be read by almost any software).

---

1. Geographic information system.

10.2 Why Use a Relational Database?

Throughout the book, we have always sided with the use of plain text, given that text files provide a simple, efficient way to store your data, code, and manuscripts. When you have much data, however, using relational databases can greatly improve performance. In this section, we discuss the main advantages of using relational databases.

Storage and Size

Databases are superior to "flat" text files when we want to store and manage large data sets. First, in a relational database, each column contains data of a specific type, and the software stores its values in the most efficient way. These special data formats not only save disk space, but can also be read into memory without any conversion. Typically, the difference between an optimal and a suboptimal solution for data storage is small, so that the effect is negligible when you are dealing with just a few thousand data points. However, suppose you want to store one billion data points—then even a small difference matters.

    The other main difference between storage of data in a text file vs. a database has to do with redundancy. Suppose that for a project, multiple people are taking measurements of samples that were treated in different ways. If you want to store all your data in a single CSV file, then for each measurement you would need to include information on the sample itself (e.g., height, weight), as well as information about the treatment (e.g., temperature, solvent) and the sampling efforts (e.g., date, person) In this case, you would end up with much information repeated across the different rows of your spreadsheet. In a relational database, you can easily and efficiently store these values in different tables: one table for the samples (where each record contains the information on a given sample), one for the sampling dates (who, when, method used), and a third table for the treatments, referencing the other two tables. Using a relational database minimizes (ideally, avoids completely) entering redundant information, thereby minimizing not only the amount of memory used, but also the probability of making mistakes during data entry.

Indexing

The speed and efficiency of retrieving data from a relational database is largely due to a special indexing system. In a text file, we can order data according to one column only (e.g., order species by name *or* by body size). If we seek data from another (unsorted) column, we would need to read the entire file from beginning to end to make sure we have found all the entries matching our search. If the data set is large, this can take a very long time.

Databases make use of an indexing system that speeds up data retrieval operations. The *index* contains one or more columns in order, and a pointer that connects the entry in the index with the appropriate location of the data in the database. If we now seek an entry from a field for which an index exists, we can find the desired value much faster because our search is performed on an ordered column.

The downside of an index is that it takes additional storage space, and it needs to be updated or rebuilt every time the data are modified. Thus, one should create indices only for fields that are frequently used for searches.

Integrity and Security

In a relational database, you can specify conditions that need to be satisfied whenever a new record is added. For example, you can reject attempts to add data whenever some fields are missing, or if certain values do not satisfy given requirements (e.g., body size needs to be positive). Moreover, the software automatically checks that a record is successfully added, and will abort changes that are not successful, returning an error without affecting the original data.

When dealing with sensitive data (e.g., human subjects), you might want to have different privileges for different users: some users can only read the data, but not modify it, others can see only partial records, others can see only summaries, etc. Most database software ships with sophisticated ways to set up privileges.

Moreover, most databases implement models of concurrency: many users can access, modify, and add data at the same time.

Client–server

Finally, when you have really large data sets, it is inefficient to have all the collaborators download the entire data. Most database systems are "client–server": the data are stored in a single machine, and the users interface with this machine through the network so that they do not need to store the data locally.

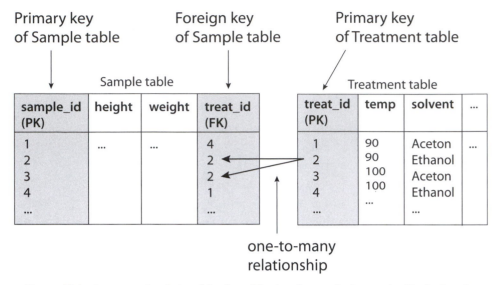

Figure 10.1. Structure of a relational database. The data for samples is organized in the Sample table. Information about the experimental conditions is organized in the Treatment table. The two tables are linked by their keys. The *primary key* of the Treatment table (on the right-hand side of the figure) becomes the *foreign key* in the Sample table, resulting in a one-to-many relationship.

Disadvantages

The downside of using databases is that data are stored in particular formats, so that you need specific software to access it. However, section 10.7.3 introduces an easy way to convert a database into a text file that can be modified in any text editor and kept under version control (assuming a reasonable file size of the resulting text file).

10.3 Structure of Relational Databases

As introduced above, data are stored in *tables*, where the rows are *records* and the columns are *fields*. For instance, assume a database with two tables as depicted in Figure 10.1. One table contains sample data with replicates for each experimental condition (Sample table). The other table contains information for each treatment (Treatment table). Each record in the Sample table is associated with a unique sample_id. This identifier is also called the *primary key* (PK) of the table. Likewise, each unique treatment in the Treatment table receives a primary key (treat_id).

Data within tables can be linked by their keys: Each sample received a specific treatment. Instead of repeating the specifications of the treatments over and over for every sample that received the same treatment, we simply refer to the treat_id. This greatly reduces redundancy. The treatment identifier in the Sample table is also called a *foreign key*. Many samples were treated in the same way. This means each record in the Treatment table is associated with many records in the Sample table (one-to-many relationship). Using primary and foreign keys, one can build one-to-one, one-to-many, and many-to-many relationships between the records in different tables.

10.4 Relational Database Management Systems

Many relational database management systems are available today. Among the free software, MySQL and PostgreSQL are the most popular and powerful. ORACLE, Informix, and SQL Server are the most diffused commercial systems. If you are using the Microsoft Office Suite, you may have noticed Access, a database manager with a simple graphical interface. Each RDBMS offers either a graphical interface maintained by the developers or many choices of free software from other providers.

In this chapter, we show examples of relational databases for biological research, using one of the simplest RDBMSs available today, SQLite.

Most RDBMSs use a client–server system that requires a complex setup. In contrast, SQLite is a serverless, zero-configuration database system and is therefore much easier to install. SQLite is a database in a single file and was originally developed to be shipped on guided missiles(!). It is now the most deployed database system: every iPhone or Android phone uses it to track apps, Firefox, Chrome, and Safari ship a copy of it, Python uses it to manage packages, and Dropbox and many other applications use it too.

10.4.1 Installing SQLite

Though it is likely that SQLite is already installed on your computer, you might not be able to access all its functionalities. The files in sql/ installation guide you step by step through the installation process.

10.4.2 Running the SQLite RDBMS

While you may prefer a graphical user interface specific to your operating system, we show examples using a simple command-line utility that is available for all platforms. Navigate to CSB/sql/sandbox and invoke the software by typing sqlite3 in your terminal:

```
$ sqlite3
SQLite version 3.13.0 2016-05-18 10:57:30
Enter ".help" for usage hints.
Connected to a transient in-memory database.
Use ".open FILENAME" to reopen on a persistent database.
sqlite>
```

The session starts by providing information on the version of SQLite, how to get help, and how to open a file. Given that we did not open an existing database, we are currently using a temporary database that will be deleted when we close the session. The prompt sqlite> indicates that the terminal is ready for your input. You can close a session by pressing Ctrl+D and interrupt execution of a command by pressing Ctrl+C.

10.5 Getting Started with SQLite

Structured Query Language (SQL) is a language for performing operations on relational databases. Though each database management system implements a particular "dialect" of SQL, most commands work in about the same way. In this book, we are going to explore SQL using SQLite.

10.5.1 Comments

There are two types of comments in SQL:

```
sqlite> -- two dashes mark a single-line comment
sqlite> /*
   ...> A backslash and asterisk mark the beginning
   ...> and the end of a comment spanning multiple lines.
   ...> */
```

10.5.2 Data Types

In a relational database (as in a data frame in R), each field contains a specific type of data. SQLite offers only a handful of data types:

NULL	Null value
INTEGER	Signed (±) integer, taking up to 8 bytes
REAL	Floating-point value, stored using 8 bytes for precision
TEXT	Strings, supporting different encodings, like UTF-8
BLOB	Binary format used to store files (e.g., images, documents) or other types

Note that SQLite does not provide two basic types of data typically found in more sophisticated databases: Boolean (TRUE/FALSE) and dates. Boolean values can be stored as integers (using 1 for TRUE and 0 for FALSE). Dates can be stored as strings `YYYY-MM-DD HH:MM:SS.SSS`, and are recognized by SQLite time and date functions.

10.5.3 Creating and Importing Tables

First of all, we need a database. We could build a database from scratch, but often we already have data in another format, for example in a `.csv` file. We will generate a database from a file that contains the results of experiments on *Daphnia* performed by Lohr and Haag (2015). The data consist of life-history traits for more than 2500 individuals, and include `ID`, the identifier for the individual; `clone`, the identifier of the isofemale line the individual belongs to; `size`, the size of the pond where the original organism starting the isofemale line was sampled ("Small" or "Large"); `pop`, the name of the population; `ad`, age at death (in days); `afr`, age at first reproduction; and `ro`, reproductive output.

We import the data into SQLite, and build the database `lohr`. To do so, launch SQLite by opening a terminal, going to the `sql/sandbox` directory, and typing `sqlite3`.

Once the SQLite session starts, we use special *dot-commands* to create the database `lohr.db`.

Dot-commands are interpreted by the `sqlite3` utility (not the underlying RDBMS) and are typically used to change the output format of queries and to import or export databases.[2]

To create our database, we set SQLite to CSV mode, import the data by indicating the location of our data file, name our table, save the database, and exit SQLite:

---

2. Type `.help` to receive an overview of all existing dot-commands. Note that dot-commands cannot have a comment in the same line.

```
sqlite> .mode csv
sqlite>  -- import CSV as lohr
sqlite> .import ../data/Lohr2015_data.csv lohr
sqlite>  -- save data as database and exit program
sqlite> .save lohr.db
sqlite> .exit
```

The .import command creates the table lohr in the computer's memory. If the table did not previously exist, the .import command automatically interprets the first row of the data as the header. However, if a table already exists in your SQLite session, all rows will be read as data. The .save command saves the database to a file on your hard disk. As such, after exiting SQLite, you should find the file lohr.db in your sandbox.

To open the file using SQLite, open a terminal and type

```
$ sqlite3 lohr.db
```

10.5.4 Basic Queries

Now that we have a database, we can start building SQL queries. But first, let's take a look around. To list the available tables, type

```
sqlite> .tables
lohr
```

Similarly, the command .schema shows the structure of the table(s):

```
sqlite> .schema
CREATE TABLE lohr(
  "ID" TEXT,
  "clone" TEXT,
  "size" TEXT,
  "pop" TEXT,
  "ad" TEXT,
  "afr" TEXT,
  "ro" TEXT
);
```

Note that all the columns in the .csv file have been imported as TEXT, though we know that some of the values are numeric. This is usually not a problem—we will see later that the data type can be specified directly in a query.[3]

Subsetting Data Using SELECT

We are now ready to write our first query. By convention, all SQL keywords are capitalized, to make them easier to distinguish from the names of the fields. Type

```
sqlite> SELECT * FROM lohr LIMIT 4;
2|1|Large|AST|87|13|236
3|1|Large|AST|99|13|200
5|1|Large|AST|102|11|250
6|1|Large|AST|72|12|215
```

You start the query with SELECT, followed by "what" you want to select (* means all the fields) and a clause specifying the table, FROM lohr. The instruction LIMIT 4 simply states that only the first four records should be returned (similar to the command head in Unix and R). Every SQLite command terminates with a semicolon.

To make the results look nicer on screen, we can use dot-commands to adjust the output format:[4]

```
sqlite> .mode column
sqlite> .header on
sqlite> .width 5 5 5 4 4 4 4
sqlite> SELECT * FROM lohr LIMIT 4;
ID    clone size  pop  ad   afr  ro
----- ----- ----- ---- ---- ---- ----
2     1     Large AST  87   13   236
```

---

3. Alternatively, you can manually create a table using the CREATE TABLE command, which requires you to specify data types. You would then read your data (e.g., .csv file) into it using the .import command. Remember to delete row headers in your .csv file as the .import command will read the entire file as data into your predefined table.

4. If you close your SQLite session, remember to call these options again so the results on your screen are formatted similarly to the output printed in the book.

```
3      1         Large  AST   99    13    200
5      1         Large  AST   102   11    250
6      1         Large  AST   72    12    215
```

We have specified the output to be formatted in columns, including a header, and set the column widths to be 5 or 4 characters long. Unfortunately, column headers are automatically cut to that length, making short (but less informative) header names preferable.

In the previous example, we used the * to select all fields in the table. To select only specific fields, list their names and separate them using commas:

```
sqlite> SELECT size, pop FROM lohr LIMIT 5;
size   pop
-----  -----
Large  AST
Large  AST
Large  AST
Large  AST
Large  AST
```

If we want to select only unique values in a field (i.e., no duplicates), add DISTINCT in front of the field names:

```
sqlite> SELECT DISTINCT size FROM lohr;
size
------
Large
Small
```

```
sqlite> SELECT DISTINCT pop FROM lohr;
pop
-----
AST
WTE
MOS
...
```

We can order the results using ORDER BY:

```
sqlite> SELECT DISTINCT pop
   ...> FROM lohr ORDER BY pop ASC;
pop
-----
AST
AST5
BEOM
BOL
ISM
```

The option ASC (ascending order) is the default behavior, and can be omitted. For descending order, use DESC.

Note that, because we have built the table using only TEXT fields, the ordering does not work well when we're dealing with numbers. To demonstrate that the alphabetical order does not match the numerical order, we want to display entries 10 to 13. We add the options LIMIT m and OFFSET n to an ORDER BY query. These options mean "return m entries after skipping n entries":

```
sqlite> SELECT DISTINCT ro
   ...> FROM lohr ORDER BY ro LIMIT 3 OFFSET 10;
ro
-----
109
11
110
```

The output shows the selected data in alphabetical order, which is not useful in most instances.

In order to sort numerically, we can convert a field of type TEXT to INTEGER directly in our query:

```
sqlite> SELECT DISTINCT ro
   ...> FROM lohr ORDER BY CAST(ro AS INTEGER) DESC
   ...> LIMIT 3;
ro
-----
```

```
493
451
440
```

Note that the CAST operation changes the data type of ro only in the output that we obtained with the SELECT operation. The data in the database lohr are not changed.

Intermezzo 10.1
  (a) Find the smallest age at first reproduction (afr) in the data.
  (b) To which population does the individual with the lowest reproductive output (ro) belong?

Filtering Using WHERE

Now that we are more familiar with basic SQL syntax, we can start building more complex queries. For example, we can use the clause WHERE to filter the records according to a criterion. Suppose we want to extract a few fields for the individuals who originate in populations sampled in Large ponds:

```
sqlite> SELECT clone, pop, size
   ...> FROM lohr
   ...> WHERE size = "Large"
   ...> LIMIT 5;
clone  pop    size
-----  -----  -----
1      AST    Large
1      AST    Large
1      AST    Large
1      AST    Large
1      AST    Large
```

You can also use greater than and less than signs (if we set the type as numeric):

```
sqlite> SELECT ro FROM lohr
   ...> WHERE CAST(ro AS INTEGER) > 140
   ...> AND CAST(ro AS INTEGER) < 142;
ro
```

```
-----
141
141
141
141
...
```

It is possible to define new names for the fields, or to create a new field by casting its type. For example, the query we just typed could have been clearer if written as

```
sqlite> SELECT CAST(ro AS INTEGER) AS ronum FROM lohr
   ...> WHERE ronum > 140 AND ronum < 142;
ronum
-----
141
141
141
141
...
```

Again, we are changing the field name only in our output, not in the underlying data.

Sometimes, we need to match records using wildcards or regular expressions. SQLite offers two ways to accomplish this:

```
sqlite> SELECT DISTINCT pop FROM lohr
   ...> WHERE pop LIKE "%T";
pop
-----
AST

sqlite> SELECT DISTINCT pop FROM lohr
   ...> WHERE pop LIKE "B%";
pop
-----
BEOM
BOL
```

```
sqlite> SELECT DISTINCT pop FROM lohr
   ...> WHERE pop LIKE "A___";  -- three underscores
pop
-----
AST5
```

The percentage sign % stands for "match zero or more characters," and the underscore _ stands for "match a single character."

The clause GLOB works as LIKE, but is case sensitive, and uses the wildcards used by Unix (see section 1.6.4):

```
sqlite> SELECT DISTINCT pop FROM lohr
   ...> WHERE pop GLOB "K*";
pop
-----
KOR
KMG
K10

sqlite> SELECT DISTINCT pop FROM lohr
   ...> WHERE pop GLOB "?1?";
pop
-----
K10
```

Operations on Groups: GROUP BY

One of the most useful functions in databases is the ability to group the data according to some criterion, and apply functions to the data once divided into groups (as with dplyr in section 9.4.2). For example, we want to know what the average life spans of individuals coming from "Large" and "Small" ponds are:

```
sqlite> SELECT size, AVG(CAST(ad AS INTEGER)) AS
     ↳ avglifespan
   ...> FROM lohr
   ...> GROUP BY size;
```

```
size    avgli
-----   -----
Large   61.99
Small   47.73
```

We have taken the field `size`, created a new field called `avglifespan` containing the average `avg` of the field `ad`, once converted into `INTEGER` and calculated for each group. The special clause `GROUP BY` is used to define the groups we want to use to perform the operation on.

Now let's compute the average life span for each `pop`:

```
sqlite> SELECT pop, AVG(CAST(ad AS INTEGER)) AS avglifespan
   ...> FROM lohr
   ...> GROUP BY pop;
pop     avgli
-----   -----
AST     61.97
AST5    88.03
BEOM    91.34
BOL     50.08
ISM     43.28
...
```

The function `AVG` is only one of many "aggregate" functions one can use in SQLite.[5] For example, the function `COUNT` is used to determine the number of records in each group:

```
sqlite> SELECT pop, COUNT(pop) AS nind
   ...> FROM lohr
   ...> GROUP BY pop;
pop     nind
-----   -----
AST     241
AST5    95
BEOM    76
BOL     85
```

5. See sqlite.org/lang_aggfunc.html for a complete list.

```
ISM    231
ISM12  71
...
```

You can filter the grouped data using the clause HAVING:

```
sqlite> SELECT pop, COUNT(pop) AS nind
   ...> FROM lohr
   ...> GROUP BY pop
   ...> HAVING nind > 200
   ...> ORDER BY nind DESC;
pop    nind
-----  -----
WTE    260
KMG    249
NFN    244
AST    241
KOR    234
ISM    231
MOS    231
VRI    224
```

Intermezzo 10.2

(a) How many distinct pops were considered?

(b) Compute the average age at first reproduction for each pop.

(c) Count how many individuals in each pop lived for more than 55 days.

Now that our SELECT queries have become more complex, you may have noticed that the order of operations matters. We recommend printing the flow chart of possible operations that is provided by the SQLite documentation,[6] and keeping it close by to avoid syntax errors.

10.6 Designing Databases

So far we have been working with a database that contains a single table. In most instances, however, you will organize your data into several tables to avoid redundancy of information. The design of a relational database is a

---

6. computingskillsforbiologists.com/sqlselect.

complex task, and you should think long and hard before starting to implement your design. Ideally, you want to split the data into logically consistent tables. The guiding principle should be *normalization*: minimize redundancy and dependency. To achieve a high degree of normalization, follow these three rules:

1. Avoid duplicate fields within a table—each field should contain unique information.
2. Each field in a table should depend on the primary key. This means that a table should contain a logically consistent set of data.
3. Tables cannot contain duplicate information. If two tables need a common field, the field should go in a third, separate table and be referenced using keys.

The concept of normalization might be easier to grasp with an example: Assume you are collecting insects at three different locations, three times a year, over the span of three years. For each day/location of sampling, you record

site data:	site of the sampling, its geographic coordinates, description of the site;
sampling data:	day, month, and year, weather conditions, temperature, humidity, etc.;
species data:	classification according to species and stage of development of all the insects collected, their total, and their measurements.

Organizing all the data in a single spreadsheet would entail a lot of repetition: the information on each site would appear several times; the information on each species would be repeated in multiple rows; each date would be represented multiple times.

Here is an alternative design that follows the rules of normalization and organizes the information into four tables, represented in figure 10.2:

Site table	Each record represents a site, and is indexed by a unique `site_id`. The fields of the record contain the name of the site, the coordinates, etc.
Sampling table	Each record is indexed by the `sampling_id` and contains information on the date, weather conditions, etc. The column `site_id` is a foreign key, associating each sampling record with the corresponding record in the site table.

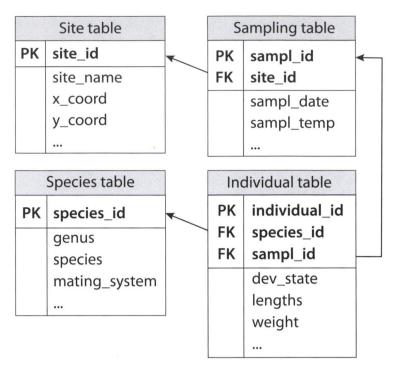

Figure 10.2. Schema of a relational database. Shown are the corresponding fields of the site, sampling, species, and individual tables. The relationship among the tables is established by their *primary keys* (PK) and *foreign keys* (FK).

Species table   Each record, indexed by `species_id`, contains information on the species that will not vary from individual to individual, such as taxonomy etc.

Individual table   Each record is indexed by the `individual_id`, and contains information on the stage of development and on measures of length, weight, etc. Each record is connected to the other tables using foreign keys: `species_id` to determine the species, `sampling_id` to connect with the sampling date. Through this latter connection, you can connect indirectly to the site table as well.

Though this organization seems more complex, it minimizes redundancy and facilitates producing counts and averages for a given site, date, species, etc.

Note that, by convention, tables should be named using an informative, singular noun.

10.7 Working with Databases

10.7.1 Joining Tables

When working with multiple tables, it will sometimes be necessary to *join* tables using their keys. To explore the use of multiple tables, we use a simplified version of the database published by Comte et al. (2016), which you can find in data/Comte2016.db. Let's load the database and take a look at its schema:

```
sqlite> .open ../data/Comte2016.db   -- run from sql/sandbox
sqlite> .schema
CREATE TABLE trans(
  "SiteID" INTEGER,
  "Per1" TEXT,
  "Per2" TEXT,
  "anguilla" TEXT,
  "lucius" TEXT
);
CREATE TABLE site(
  "SiteID" INTEGER PRIMARY KEY,
  "CoordX" REAL,
  "CoordY" REAL,
  "Basin" INTEGER,
  "Urban" REAL,
  "Fragm" REAL
);
```

There are two tables in the database: trans describes the transitions in the state of two populations (anguilla, eel; lucius, pike) from a certain time point (Per1) to another (Per2) at a given location (SiteID). The specifics of the location are stored in the table site, which contains the GPS coordinates, a code specifying the river basin the site belongs to (Basin), as well as measures of urbanization (Urban) and fragmentation (Fragm). The field SiteID is a *primary key* of the table site (i.e., each record has a unique value), and a *foreign key* of the table trans (i.e., many records have the same SiteID). This means that there is a one-site → many-transitions relationship between the records.

The operation of combining tables is called a *join*. Only a few types of join are available in SQLite.[7] The *inner join* is the most common and is therefore the default in SQLite (i.e., specifying INNER is optional in your query). In an inner join, only records that have a matching key in both tables will be joined (see figure 10.3). In contrast, an *outer join* will include all the records that are matched in both tables, as well as those that are present in the table provided in the FROM clause. The fields that have no corresponding value in the other table will be set to NULL.

We will join the tables trans and site on the key SiteID to illustrate the basic syntax of a join statement:[8]

```
sqlite> .mode line
sqlite> SELECT * FROM trans INNER JOIN site
   ...> ON trans.SiteID = site.SiteID LIMIT 1;
   SiteID = 27900
     Per1 = P1
     Per2 = P2
 anguilla = persisted
   lucius = unoccupied
   SiteID = 27900
   CoordX = 522459.14
   CoordY = 2142197.27
    Basin = 4
    Urban = -0.47
    Fragm = 0.0
```

We obtain a table containing the fields of both tables by pairing each record in trans with the corresponding record in site. This returns a very wide table, so for clarity, we set the output mode to line, which shows the fields one after the other, instead of printing them next to each other. Alternatively, you can choose the mode column and widen your terminal window to see the full joint table. Take special care when setting up join statements, as they can produce results with very many fields and records.

---

7. Other flavors of SQL provide additional types of joins. More on joins in SQLite can be found at computingskillsforbiologists.com/sqljoins.

8. In our example database, an inner join and outer join produce the same result since all records have a corresponding value in the other table.

INNER JOIN

```
SELECT M, TableA.F, TableB.F, N
FROM TableA
INNER JOIN TableB
ON TableA.F = TableB.F
```

LEFT (OUTER) JOIN

```
SELECT M, TableA.F, TableB.F, N
FROM TableA
LEFT JOIN TableB
ON TableA.F = TableB.F
```

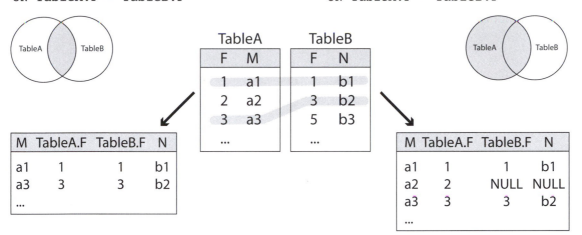

Figure 10.3. Inner and outer joins. Two tables (TableA and TableB) are joined using either an inner join (left-hand side of the image) or an outer join (right-hand side of image). The output of the inner join will include only records that are present in both tables. The outer join will also include records that are present in the table that was specified in the FROM clause. Hence, it matters whether you join A to B or B to A.

Suppose we want a count of the types of transitions of anguilla from periods P1 to P2, across sites:

```
sqlite> .mode column
sqlite> SELECT anguilla, COUNT(anguilla) AS num
   ...> FROM trans WHERE Per1 = "P1" AND Per2 = "P2"
   ...> GROUP BY anguilla;
anguilla     num
----------   ----------
persisted    258
colonized    13
extirpated   21
unoccupied   150
```

Now we would like to see the same data, grouped by river basin (i.e., the field Basin in the site table). We can achieve this only by joining the two tables:

```
sqlite> SELECT Basin, anguilla, COUNT(anguilla)
   ...> FROM site INNER JOIN trans
   ...> ON site.SiteID = trans.SiteID
   ...> WHERE Per1 = "P1" AND Per2 = "P2"
   ...> GROUP BY Basin, anguilla;
Basin       anguilla    COUNT(anguilla)
----------  ----------  ---------------
1           persisted   12
1           unoccupied  4
2           persisted   26
2           colonized   3
2           extirpated  7
2           unoccupied  32
...
```

10.7.2 Views

In many cases, writing a complex query is tedious and time consuming, and we would like to store the results to be used later. A *view* is a virtual table that we can query exactly as if it were a proper table, but it does not occupy any space on disk—it is generated on the fly when needed. Creating a view is very easy, as it is sufficient to put the clause CREATE VIEW myname AS in front of the query.

For example, create a view called both for the join of the two tables in Comte2016:

```
sqlite> CREATE VIEW both AS
   ...> SELECT * FROM site INNER JOIN trans
   ...> ON site.SiteID = trans.SiteID;
```

All views are listed exactly as the tables:

```
sqlite> .tables
both    site    trans
```

Intermezzo 10.3

   (a) Calculate the average value of urbanization (Urban) per Basin.

   (b) Find all the unique Basins where the transitions of both lucius and anguilla from P3 to P4 are classified as extirpated.

10.7.3 Backing Up and Restoring a Database

To backup a database, you can convert the entire database into a text file and even put it under version control (note that GitHub has a 1 Gb size quota).

First, specify the location and file name of the database backup, then *dump* the data and (optionally) close the database:

```
sqlite> .output [PATHTOFILE]/sqlitedumpfile.sql
sqlite> .dump
sqlite> .exit
```

The .sql ending of the output file name is arbitrary but provides a hint to what the file contains. Have a look at the sqlitedumpfile.sql in your terminal and you will see that the text file contains SQL statements to rebuild the database.

```
$ head sqlitedumpfile.sql
PRAGMA foreign_keys=OFF;
BEGIN TRANSACTION;
CREATE TABLE trans(
  "SiteID" INTEGER,
  "Per1" TEXT,
  "Per2" TEXT,
  "anguilla" TEXT,
  "lucius" TEXT
);
INSERT INTO "trans" VALUES(27900,'P1','P2','persisted','
    ↳ unoccupied');
```

There are two options to restore a database, either directly on the command line or from within SQLite:

```
# restore database on command line
$ sqlite3 my_database.db < sqlitedumpfile.sql
# open database to restore from within SQLite
$ sqlite3 my_database.db
sqlite3> .read sqlitedumpfile.sql
```

Note that our sqlitedumpfile.sql file contains CREATE TABLE commands. The restore will therefore fail if tables with those names are already present in the database. Either start with a new database (following the example above) or drop existing tables first, as shown in section 10.7.4.

10.7.4 Inserting, Updating, and Deleting Records

One of the main features of databases is that whenever you change, add, or remove records and tables, the changes are immediately implemented, and you cannot (typically) reverse the changes. As such, you should be quite careful, and back up your data before attempting updates.

Inserting a Record into a Table

In most cases, you will import your data from a set of files, or build your database programmatically (see below). If you need to insert some records by hand, however, you can use an INSERT query:

```
sqlite> INSERT INTO mytable (field1, field2, ..., fieldN)
   ...> VALUES ("abc", 23, ..., "March_15")
```

You can use a SELECT query to extract the VALUES from another table:

```
sqlite> INSERT INTO mytable
   ...> SELECT col1, col2, ..., colN FROM othertable;
```

Updating Records

Sometimes you need to update a certain set of records. In these cases, you want to build an UPDATE query. For example, if you want to change all the records having anguilla = "persisted" to anguilla = "Persisted" in the trans table, you would run this command:

```
sqlite> UPDATE trans SET anguilla = "persisted"
   ...> WHERE anguilla = "Persisted";
```

Deleting Records

The syntax is similar to that of UPDATE:

```
sqlite> DELETE FROM mytable
   ...> WHERE mycondition;
```

Note that a DELETE query without the WHERE clause will delete all records from your table.

Deleting Tables and Views

To delete a table, use the command

```
sqlite> DROP TABLE mytable;
```

For views use

```
sqlite> DROP VIEW myview;
```

Be careful with these operations, as there is no way to recover the data once you have deleted it (unless you have a backup, of course!).

10.7.5 Exporting Tables and Views

While SQLite by default sends the output to your terminal, you can redirect the output to a file. For example,

```
sqlite> .mode list
sqlite> .separator ,
sqlite> .output test.csv
```

```
sqlite> SELECT DISTINCT Basin, SiteID FROM site;
sqlite> .output
```

The second .output resets the output redirection, so that you can see the results again in the terminal.

10.8 Scripting

You can create scripts to be run by SQLite. Simply collect into a text file (with extension .sql) all the commands you would type. Then execute the commands in the script by invoking SQLite from the command line:

```
$ sqlite3 my_script.sql
```

Given that these scripts are regular text files, you can put them under version control.

10.9 Graphical User Interfaces (GUIs)

At first, it can be intimidating not to have an easy way to see the data and to have to explore it in a terminal. To ease the transition, you might want to choose a program with a graphical user interface that helps you manage your databases and build your queries. For example, check out DB Browser for SQLite[9] and SQLite Studio,[10] among many others.

10.10 Accessing Databases Programmatically

You can use databases in your programs, using for instance Python or R. Both give you access to different types of databases, including SQLite. In general, you will need to *connect* to a database, and create a *cursor* that you can use to execute queries, create tables and views, etc. The SQL query itself is formed in the same way we have seen so far, without the semicolon that usually concludes an SQL statement.

---

9. sqlitebrowser.org.
10. sqlitestudio.pl.

10.10.1 In Python

The following Python script shows you how to connect to an SQLite database:

```python
#!/usr/bin/python
import sqlite3 # library installed by default

# create a connection to the database
con = sqlite3.connect('../data/Comte2016.db')
# create a cursor
cursor = con.execute("SELECT DISTINCT Basin FROM site")
for row in cursor:
    print("Basin ", row[0])
# before exiting the program, close the connection
con.close()
```

Furthermore, the Python library `sqlalchemy` integrates SQLite and pandas for convenient data manipulation.[11]

10.10.2 In R

The same can be accomplished in R by using the `RSQLite` package:

```r
1   # install the package if needed
    install.packages("RSQLite")
    library(RSQLite) # load the package
4   # specify the type of database
    sqlite <- dbDriver("SQLite")
    # connect to the database
7   con <- dbConnect(sqlite, "../data/Comte2016.db")
    # run a query and assign to data frame
    results <- dbGetQuery(con, "SELECT DISTINCT Basin FROM
10  site")
    print(results)
    # close connection
13  dbDisconnect(con)
```

---

11. computingskillsforbiologists.com/sqlalchemy.

There are also options to retrieve really large data sets in batches.[12]

10.11 Exercises

10.11.1 Species Richness of Birds in Wetlands

Żmihorski et al. (2016) reported the number of birds in several wetlands in Sweden over the course of four years. The data are contained in Zmihorski2016.db, and the description of the fields in the table tbirds is reported in Zmihorski2016_about.txt.

1. Each Area contains several sites (PointID). Calculate the average Species.richness for each site (put the field name within quotes as SQLite does not like field names containing punctuation!).
2. Calculate the average Species.richness for each Area/Year combination.
3. Calculate whether more species are found (on average) when the area is flooded (field Flooding.binary), or not flooded.
4. [**Advanced**] Produce a count of the co-occurrence of three bird species (yellow wagtail, YW; lapwing, LW; skylark, SK), producing a table like

YW	LW	SK	Occurrences
0	0	0	58
0	0	1	18
0	1	0	24
0	1	1	15
1	0	0	80
1	0	1	59
1	1	0	5
1	1	1	204

10.11.2 Gut Microbiome of Termites

Mikaelyan et al. (2015) studied the gut microbial communities of higher termites, which feed on a variety of food sources besides wood. They found that the diet of termites strongly influences the bacterial community of the gut.

---

12. computingskillsforbiologists.com/sqliteinr.

The database Mikaelyan2015.db contains three tables: table tSpp contains the names of the termite species, tOTU contains the code for the observable taxonomic unit of bacteria (i.e., the bacterial "species"), and table tNumber contains the de-replicated frequency of occurrence of a certain OTU in the gut of a given species.

1. By joining the three tables, create a view displaying Spp (species name), OTU (code for the OTU), and Num, the number of occurrences of the OTU in the gut of the species. Notice that Num can be (and often is) 0.
2. Count the number of OTUs found in each species (remember to filter using Num > 0).
3. List the OTUs that are found in four or more species.
4. Delete all the records in tNumber where Num is 0.
5. [**Advanced**] Using the package vegan in R, compute Shannon's index of diversity for the gut microbial community of each species.

10.12 References and Reading

A list of books on SQLite:
   sqlite.org/books.html.

A free book on SQL from the "Learn X The Hard Way" series:
   computingskillsforbiologists.com/sqlthehardway.

The tutorialspoint.com website has a fantastic collection of short pages on SQLite commands:
   computingskillsforbiologists.com/learnsqlite.

SQLite in Python:
   computingskillsforbiologists.com/pythonsqlite.

SQLite in R:
   computingskillsforbiologists.com/rsqlite.

CHAPTER 11

• • • • • • • • • • • •

Wrapping Up

We have presented a variety of computational tools that can be used in concert to build complex pipelines for the analysis and presentation of your biological data. We have emphasized breadth over depth, and showed how each tool was born to address a certain set of problems, so that picking the right tool for each task will make your life easier.

The strength of these tools is that they can be seamlessly integrated, allowing you to automate much of your work. For example, you can extract the relevant data with a few Unix commands, send the data to Python for analysis, and finally call R to perform statistics and draw figures. Switch freely between tools to achieve a certain goal, but keep in mind the advantages and disadvantages of each tool in order to make an informed choice. As the adage goes, "If all you have is a hammer, everything looks like a nail"—we strongly believe that a good working knowledge of many tools will make you more productive than mastering just a single tool.

Clearly, a few pages, and a few hours spent on the exercises will not give you the level of proficiency needed to solve all of your research problems. However, the material presented in this book should provide you with an excellent starting point so that you can refine your skills by practicing them in your daily research.

It is important to practice regularly, and protect time to play with new tools. While it is great to have a specific task in mind when starting to program, it is also important to leave time to acquire and improve new skills. You will be happier if you think "I have invested two hours in learning Python!" rather than "It took me two hours to solve this easy problem in Python— I could have done it in Excel in 20 minutes!" While this might be true when you start, your investment will pay back in the long run. With a bit of perseverance, you will start reusing code to automate processes, thereby saving a lot of time. Even better, you will tackle problems that you could not have addressed at all without your new skills.

11.1 How to Be a More Efficient Computational Biologist

This section provides additional, practical suggestions on how to work with computers in biology. Following it will likely avert major catastrophes.

Back Up

Needless to say, it is very important to back up your data, code, and manuscripts frequently. An online version control system for all your text files adds an extra layer of safety.

Data Are Sacred

In your programs, never alter the original data you have collected/collated! Read the data into memory, or create a temporary copy, and modify that. We have seen too many mistakes resulting from ordering all but one column in your favorite spreadsheet software. The data, once collected, should never be changed. The same goes for the metadata.

Divide et Impera

Divide a project into subprojects that are easy to solve. Never rely on a single, long function to carry out the whole analysis from start to finish. Rather, divide the task into logical steps, and tackle each step separately. This will help you recycle lots of code, besides being a more efficient way of working.

Test It on Paper

When starting a complex program, try to work out the flow of the code on a piece of paper/blackboard before you start. Even better, explain what you're going to do to a friend or colleague. This will force you to consider aspects of your code that are easily overlooked.

Do It Well

There's no point in writing "quick and dirty" code. In fact, that's when most mistakes sneak in. Plan a program with pseudocode, and write it as well as you can. Fix and improve your programs with care if you find bugs at a later point or if you find a way to write it more efficiently. Avoid extemporaneous code to hack/fix your programs—almost surely, you're introducing a bug.

Write to Be Read

We have stressed this point quite a few times, and we reiterate it once again: write your code so that other people can read it and understand it (including yourself in six months).

Plan Ahead, Do a Little at a Time

Coding, and especially debugging, can be boring. Find a good time to do it, and break it into small units to stay motivated and focused. Plan, code, test, repeat!

Automate

Computers err in systematic ways, but you don't. It is easier to fix systematic mistakes than random ones that are introduced in "manual" operations. If you think(!) it takes you X hours to do it "by hand," and $3X$ hours to automate it, then automate it. There is a high likelihood that you will have to repeat a task more often than you think and you will thank yourself for automating it the first time around.

Keep a Log

Find a place to store your aha moments, snippets of code or useful commands, instructions on how to do a certain thing. You probably keep an extensive lab book if you are conducting experimental work. Get into the same habit when it comes to programming and keep a "computational analysis lab book." Digital or paper, write down your reasoning and ideas. These things are easy to forget, and keeping good notes will help you in the future. For example, gist.github.com provides a way to store snippets of code and notes.

11.2 What Next?

If you found this material useful, and you want to further your knowledge of scientific computing, you might want to consider the following ideas, techniques, and tools that can make you a stronger, more productive, and ultimately happier scientist.

Master Your Text Editor

Text editors and integrated development environments (IDEs) are primarily meant for writing code, but they can do much more for you! Most of them can perform neat tricks, such as matching parentheses, name completion, editing many files at once, search using regular expressions. Mastering your editor will help you code faster and with fewer mistakes due to typos etc.[1]

Packages

As we have tried to convey in their respective chapters, R and Python can be extended through the use of packages and modules.[2] Many of these packages are domain specific: try to find those that are most useful for your research, study them in detail, and possibly extend them (that's the beauty of open source).

Parallel/Cluster Computing

Some of your code will take a long time to run (either because you are performing complex operations, or because you are iterating over much data). In many cases, you can divide the problem into subproblems, each of which can be run independently (or semi-independently) from the rest. In these cases, you can exploit the fact that your computer is multicore (i.e., has several processors, each of which can run your code independently), or distribute the computing through a computer cluster (universities typically have at least a few of them on campus; otherwise look into Amazon Web Services or similar solutions).

Other Languages

In this book, we have introduced R and Python, both of which can be slow in some circumstances. If you need to perform many (i.e., billions or more) operations, then you need to speed things up dramatically. Rather than spending hundreds of hours hacking your R or Python code to make it 20% faster, consider using languages with great time performance, such as C or Java. Though they are a little more involved than the languages covered in this book (e.g.,

---

1. If you are undecided, check out xkcd.com/378.
2. See computingskillsforbiologists.com/rpackages for R and pypi.org for Python.

you need to define a type for each variable), you will find that they have many features in common with the languages you've just learnt.

Metaheuristics

Stochastic global optimization algorithms, aka "metaheuristics," are one set of techniques that can save the day in biological research. The goal is to find the "best" solution for problems that are too complex to be solved analytically, and where there are many nearly optimal solutions—possibly very "distant" from each other. Problems of this kind arise in many branches of biology, including protein folding, reconstruction of phylogenetic trees, and estimation of maximum-likelihood parameters for complex models, to name a few.

Interestingly, many of these techniques were inspired by biology (e.g., genetic algorithms, ant colony optimization). Understanding simulated annealing, genetic algorithms, and parallel tempering should get you started, and allow you to attack a whole new class of problems that would be difficult to solve otherwise.[3]

Open Science

Throughout the material, we have emphasized the importance of sharing code and knowledge. The Open Science movement advocates doing science "en plein air," making all the data, code, figures, and manuscripts publicly and immediately available. Creating public repositories on GitHub or Bitbucket to deposit your code and data, and sharing your manuscripts on bioRXiv, arXiv, or other systems would go a long way toward removing the barriers among scientists, and leveling the playing field of science so that more people can participate, and science progress faster (as the University of Chicago's motto says, *Crescat scientia; vita excolatur*!).

Some people resist this idea of openness because of their fear of being "scooped" (i.e., other research groups taking advantage of openly available data and manuscripts to beat the original authors to the finish line of a big discovery). In a way, publishing preprints and citable data lessens the problem, by establishing authorship earlier. Every online repository has time stamps

---

3. Liz Sander, a previous lab member, has written a nice, brief introduction to heuristic optimization, which includes Python code. You can find it at computingskillsfor biologists.com/heuristics.

showing what was committed and when. Such proof goes a long way to establishing "who was first." However, we recognize that the system of incentives in science and academia needs to be altered in order to further encourage collaboration and sharing.

11.3 Conclusion

We hope that you have found the material interesting, the exercises challenging and fun, and that you now feel more confident when confronted with computational problems.

We welcome any feedback you might have on the material: please contact us for typos, mistakes, better ways to approach this or that problem, ideas for fun exercises—any suggestions for making the material more accessible and useful. You can contact us via the form at computingskillsfor biologists.com/contact. We truly care about your comments, and will amend the material accordingly.

Happy computing!

Intermezzo Solutions

• • • • • • • • • • • • • • • • • •

Unix

Intermezzo 1.1

(a) `cd ~`

(b) `cd CSB/unix/sandbox`

assuming you stored CSB in your home directory.

(c) `cd ../../python/data`

(d) `cd ~/CSB/python/sandbox`

(e) `cd -`

Intermezzo 1.2

(a) `cd ~/CSB/unix/data`

(b) `wc -l Marra2014_data.fasta`

(c) `touch ../sandbox/toremove.txt`

(d) `ls ../sandbox`

(e) `rm ../sandbox/toremove.txt`

Intermezzo 1.3

(a) `cd ~/CSB/unix/data/`
`cut -d ";" -f 5 Pacifici2013_data.csv | sort | head -n 1`
`cut -d ";" -f 5 Pacifici2013_data.csv | sort | tail -n 1`

(b) `cut -d ";" -f 3 Pacifici2013_data.csv | tail -n +2 | sort |`
`uniq | wc -l`

Intermezzo 1.4

(a) ```
cd ~/CSB/unix/sandbox/
find ../../ -name "*Dalziel*.csv" -type f
cp ../../python/data/Dalziel2016_data.csv .
```
(b) ```
head Dalziel2016_data.csv
cut -d "," -f 3 Dalziel2016_data.csv | tail -n +2 | sort |
uniq
cut -d "," -f 3 Dalziel2016_data.csv | tail -n +2 | sort |
uniq -c
```
(c) ```
grep -i washington Dalziel2016_data.csv | cut -d "," -f 4 |
sort -n -r | head -n 1
```
(d) ```
tail -n +2 Dalziel2016_data.csv | sort -t "," -k 4 -n -r |
head -n 1
```

Version Control

Intermezzo 2.1

(a) ```
echo "June 18th, 1858: read essay from Wallace" > todo.txt
```
(b) ```
git add todo.txt
```
(c) ```
git commit -m "Added to-do list"
```

*Intermezzo 2.2*

(a) ```
cd ~/CSB/git/sandbox
```
(b) ```
mkdir thesis
cd thesis
git init
```
(c) ```
echo "The best introduction ever" > introduction.txt
```
(d) ```
git add introduction.txt
git commit -m "Started introduction"
```
(e) ```
git branch newintro
git checkout newintro
```
(f) ```
echo "A much better introduction" > introduction.txt
touch methods.txt
git add --all
git commit -m "A new introduction and methods file"
```
(g) ```
git log --oneline --decorate
```

(h) `git checkout master`
`git merge newintro`
`ls`
`cat introduction.txt`

(i) `git branch -d newintro`
`git log --oneline --decorate`

Basic Programming

Intermezzo 3.1

(a) `s = "WHEN on board H.M.S. Beagle, as naturalist"`
(b) `s.count("b")`
(c) `s.lower().count("b")`
(d) `s.replace("WHEN","When")`

Intermezzo 3.2

(a) `a = [1, 1, 2, 3, 5, 8]`
(b) `a[4:6]`
`a[-2:]`
(c) `a.append(13)`
`a`
(d) `a.reverse()`
`a`
(e) `m = "a": ".-", "b": "-...-", "c": "-.-."`
(f) `m["d"] = "-.."`
`m`
(g) `m["b"] = "-..."`
`m`

Intermezzo 3.3

(a) `range(3, 17)` yields the numbers from 3 to 16, thus `hello` will be printed 14 times.
(b) `range(12)` returns the numbers from 0 to 11. For each number whose remainder is 0 when divided by 3 (i.e., the four numbers 0, 3, 6, and 9), `hello` is printed.
(c) `range(15)` returns the numbers from 0 to 14. For each number whose remainder is 3 when divided by 5 or 4, `hello` is printed (i.e., `hello` is printed for j = 0, 7, 8, 11, 13).

(d) For each cycle of the loop, we add 3 to z (starting with z = 0). Then hello is printed 5 times before z != 15 becomes False (i.e., before z == 15).

(e) The loop starts with z set to 12. In each cycle, we test whether z is smaller than 100 and evaluate whether z equals 31. If this latter condition is met, hello is printed 7 times. In each cycle, we also test whether z equals 18. If the condition evaluates to True, hello is printed (once). As such, hello is printed 8 times in total.

(f) range(10) returns the numbers from 0 to 9. In each cycle, we test whether i is greater than 5. As long as the if statement is False (i.e., i has values 0 to 5), we print hello—6 times in total. When the if condition is True (i.e., when z takes the value 6), we encounter a break statement that terminates the loop.

(g) As long as z is smaller than 25, we add 1 to z at every cycle. When the if condition is met (i.e., z is odd), the continue statement is executed and skips the remainder of the cycle, so the print statement is not executed. Only when the if statement is not True (i.e., z is even) is hello printed. There are 12 even numbers between 1 and 25.

Intermezzo 3.4

```
import csv
with open ("../data/Dalziel2016_data.csv") as f:
    reader = csv.DictReader(f)
    for row in reader:
        print(row["loc"], row["pop"])
```

Writing Good Code

Intermezzo 4.1

(a) Returns $\sqrt{x}$

(b) Returns the largest of two numbers

(c) An inefficient way to sort three numbers into ascending order

(d) Calculates $x!$ (i.e., $x(x-1)(x-2)\dots 1$)

(e) Finds the prime factorization of an integer (i.e., the unique list of prime numbers that when multiplied together yield the number)

(f) Calculates $x!$ using recursion

(g) Generates a list of prime numbers between 2 and x (inclusive, assumes $x \geq 2$)

Intermezzo 4.2

The line i = i + 4 should be i = i + 3: you want to move from the first base to the fourth, and then to the seventh, etc. Remember that Python starts counting at 0.

Regular Expressions

Intermezzo 5.1

(a) Matches one digit: "2"

(b) Matches zero or more word characters ($\w*$), followed by a white space ($\s$), followed by a digit ($\d$), zero or more characters ($.*$), and ending with a digit ($\d$): "take 2 grams of H2"

(c) Any sequence of word characters (zero or more), flanked by two white spaces: " upon "

(d) A sequence of one to three word characters, flanked by two white spaces: " a "

(e) Matches the last word in the target string (preceded by a white space): " time"

Intermezzo 5.2

(a) r"[A-Za-z]{3}\d{5}"

(b) r"[A-Za-z]{4}\d{8,10}"

(c) r"([A-Z]{1}\d{5}|[A-Z]{2}\d{6})"

Scienctific Computing

Intermezzo 6.1

(a) Yes, the answer is π^4:

```
import numpy as np
import scipy.integrate
```

```
def integrand(x):
    return((scipy.log(x) ** 2) / x) * scipy.log((1 +
        ↳ x) / (1 - x))


# perform the calculation
answer = 24 * scipy.integrate.quad(integrand, 0, 1)
    ↳ [0]
# is this pi^4?
print(answer ** (1 / 4))
```

(b) The proportions cycle when $x(0) = [\frac{1}{4}, \frac{1}{4}, \frac{1}{2}]^t$, and remain constant when $x(0) = [\frac{1}{3}, \frac{1}{3}, \frac{1}{3}]^t$:

```
import numpy as np
import scipy.integrate

def fitness(x, A):
    """

    compute the fitness of all types
    given payoff matrix A
    and proportions x
    (with x[1] + x[2] + ... + x[n] = 1)
    """

    return(np.dot(A, x))

def replicator(x, t, A):
    """

    replicator dynamics
    for linear fitness
    d xi / dt = xi ((A x)i - x^t A x)
    """
    fi = fitness(x, A)
    phi = sum(x0 * fi)
    return(x * (fi - phi))

# payoff matrix
A = np.reshape([0, 1, -1,
               -1, 0, 1,
                1, -1, 0], (3, 3))
```

```
# time for integration
t = np.linspace(0,100,1000)
# initial conditions
x0 = np.array([1/4, 1/4, 1/2])
# integrate dynamics (note that extra arguments must
    ↳ be in tuples)
x = scipy.integrate.odeint(replicator, x0, t, args =
    ↳ (A,))
# plotting
get_ipython().magic(u"matplotlib inline") # plot
    ↳ inline
import matplotlib.pyplot as plt
plt.plot(t, x[:, 0], label = "x1")
plt.plot(t, x[:, 1], label = "x2")
plt.plot(t, x[:, 2], label = "x3")
plt.legend(loc = "best")
plt.xlabel("t")
plt.grid()
plt.show()
# the three types cycle up and down
```

Intermezzo 6.2

(a)
```
import pandas
# read the data
data = pandas.read_csv("../data/Gachter2016_data.csv")
# see the structure
data.head()
# extract data for Claim == 0 and copy the DataFrame
claim_0 = data[data.Claim == 0].copy()
# country with the lowest frequency of 0s [Tanzania]
print(claim_0.sort_values("CumulativeFrequency").head
    ↳ (1))
# country with the highest frequency of 0s [Germany]
print(claim_0.sort_values("CumulativeFrequency",
    ↳ ascending = False).head(1))
# extract the data for Claim == 4 and copy the
    ↳ DataFrame
```

```
claim_5 = data[data.Claim == 4].copy()
# update values
claim_5.Claim = 5
claim_5.CumulativeFrequency = 1.0 -
    ↳ claim_5.CumulativeFrequency.values
# country with the lowest frequency of 5s [Lithuania]
print(claim_5.sort_values("CumulativeFrequency").head
    ↳ (1))
# country with the highest frequency of 5s [Morocco]
print(claim_5.sort_values("CumulativeFrequency",
    ↳ ascending = False).head(1))
```

Statistical Computing

Intermezzo 8.1

(a) ```
z <- seq(from = 2, to = 100, by = 2)
print(z)
```
(b) ```
divisible_by_12 <- z[z %% 12 == 0]
print(length(divisible_by_12))   (8)
```
(c) `print(sum(z))`   (2550)
(d) `print(sum(z) == 51 * 50)`   (True)
(e) `print(z[5] * z[10] * z[15])`   (6000)
(f) ```
y <- 3 * (0:10)
print(y[y %in% z])   (6, 12, 18, 24, 30)
```
(g) `seq(2, 100, by = 2) == (1:50) * 2`

A better solution is to use `all.equal`:

`all.equal(seq(2, 100, by = 2), (1:50) * 2)`   (True)
(h) `z ^ 2`   (produces a new vector containing the elements of z squared)

### *Intermezzo 8.2*

(a) `print(mean(trees$height))`   (76)
(b) `print(mean(trees[trees$height > 75, "girth"]))`   (14.44167)
(c) `print(max(trees[(trees$volume >= 15) & (trees$volume <= 35), "height"]))`   (86)
```

Intermezzo 8.3

(a) `max(ch6$nA1A1 + ch6$nA1A2 +ch6$nA2A2)` (124)

(b) `max(rowSums(ch6[, 5:7]))` (124)

(c) `sum(rowSums(ch6[, 5:7]) == ch6$nA1A1) +`
    `sum(rowSums(ch6[, 5:7]) == ch6$nA2A2)` (i.e., we sum the SNPs in which all individuals are A1A1 or A2A2: 2683)

(d) `sum(ch6$nA1A2 < rowSums(ch6[, 5:7]) * 0.01)` (i.e., we compute for how many SNPs are less than 1% of the individuals heterozygous: 3368).

Intermezzo 8.4

(a) `if (z > 100) print(z ^ 3)`

(b) `if (z %% 17 == 0) print(sqrt(z))`

(c) `if (z < 10) print(seq(1, z))`

Intermezzo 8.5

(a) Generates a vector containing the numbers 1, 4, 7, 10, ..., 1000. Print all those that are divisible by 4.

(b) Prints z only if it is prime, but tests for primality in a very inefficient way: checks whether z is divisible by any number between 2 and $z - 1$.

Intermezzo 8.6

(a)
```
# assign label "Exp" at random
trees$Exp <- sample(c(0, 1),
                    nrow(trees),
                    replace = TRUE)
```

(b)
```
# function to randomize label "Exp"
randomize_Exp <- function(trees){
  trees$Exp <- sample(c(0, 1),
                    nrow(trees),
```

```r
                                replace = TRUE)
  return(trees)
}


# number of randomizations
n_rand <- 100
# initialize p-values
pvalues <- rep(-1, n_rand)
# repeat n_rand times
for (i in 1:n_rand){
  # randomize Exp label
  trees <- randomize_Exp(trees)
  # compute t.test
  ttest_res <- t.test(trees[trees$Exp == 0,
      ↳ "volume"],
                        trees[trees$Exp == 1,
                            ↳ "volume"])
  # store probability
  pvalues[i] <- ttest_res$p.value
}
# proportion of "significant" p-values
print(sum(pvalues < 0.05) / n_rand)
```

```r
(c) is_sum_pentagonal <- function(x1, x2, x3){
  # y is the sum of the three numbers
  y <- x1 + x2 + x3
  # store condition
  tmp <- (sqrt(24 * y + 1) + 1) / 6
  if (as.integer(tmp) == tmp) {
    return(TRUE)
  }
  return(FALSE)
}


# some tests
is_sum_pentagonal(0, 0, 1) # 1 -> TRUE
is_sum_pentagonal(2, 2, 1) # 5 -> TRUE
is_sum_pentagonal(6, 6, 1) # 13 -> FALSE
is_sum_pentagonal(6, 6, 0) # 12 -> TRUE
```

Data Wrangling and Visualization

Intermezzo 9.1

(a)
```
popsize %>% select(Year) %>%
   distinct() %>% arrange()
```

(b)
```
ndvi %>% top_n(1, NDVI_May)
```

(c)
```
# version 1
popsize %>% filter(Herd == "WAH") %>%
   arrange(desc(Pop_Size)) %>% head(3) %>%
   select(Year)

# version 2
popsize %>% filter(Herd == "WAH") %>%
   top_n(3, Pop_Size) %>%
   select(Year)
```

Intermezzo 9.2

(a)
```
popsize %>% group_by(Herd) %>%
   summarise(avgPopSize = mean(Pop_Size))
```

(b)
```
ndvi %>% group_by(Herd) %>%
   summarise(ndvi_sd = sd(NDVI_May)) %>%
   top_n(1, ndvi_sd) %>% select(Herd)
```

(c)
```
popsize %>% mutate(Relative_Pop = Pop_Size /
                        mean(Pop_Size))
```

Intermezzo 9.3

(a)
```r
# using tidy form
seaice <- read_tsv("../data/FauchaldEtAl2017/
    ↳ sea_ice.csv")
seaice <- seaice %>% gather(Month, Cover, 3:14)
seaice %>% group_by(Herd, Month) %>%
  summarise(avg_cover = mean(Cover))
```

(b)
```r
# using the original format
seaice <- read_tsv("../data/FauchaldEtAl2017/
    ↳ sea_ice.csv")
# store unique Herds
herds <- sort(unique(seaice$Herd))
# to store results
res <- data.frame()
for (h in herds){
  tmp <- seaice %>% filter(Herd == h) %>%
    select(-Herd, - Year)
  # compute mean by month
  avgmonth <- apply(tmp, 2, mean)
  # store
  res <- rbind(res, data.frame(
    Herd = rep(h, length(avgmonth)),
    Month = names(avgmonth),
    avg_cover = as.numeric(avgmonth)
  ))
}
print(res)
```

Intermezzo 9.4

(a)
```r
# compute average Pop_Size per Year/Herd
avg_popsize <- popsize %>%
  group_by(Herd, Year) %>%
  summarise(avg_ps = mean(Pop_Size))
```

```r
# compute average sea-ice cover
avg_Perc_seaicecover <- seaice %>%
  group_by(Herd, Year) %>%
  summarise(avg_cover = mean(Cover))
# join the tables
avg_Perc_seaicecover <- inner_join(avg_popsize,
                                   avg_Perc_seaicecover)
```

(b)
```r
missing_snow <- anti_join(popsize, avg_Perc_
    ↳ snowcover)
```

(c)
```r
# load snow cover
snow <- read_tsv("../data/FauchaldEtAl2017/snow.csv")
# take a peek
head(snow, 3)
# extract snow cover in March
cover_March <- seaice %>% filter(Month == "Mar")
# join the tables
cover_March <- inner_join(cover_March, snow)
# compute Kendall's correlation
cover_March %>% group_by(Herd) %>%
  transmute(tau = cor(Cover,
                      Week_snowmelt,
                      use = "pairwise.complete.obs")
                         ↳ # this removes NAs
          ) %>% # now group by Herd and tau and use
                ↳ distinct to avoid repetitions
  group_by(Herd, tau) %>% distinct()
```

Intermezzo 9.5

(a)
```r
# explore data snow
# create scatter plot with continuous variables
ggplot(data = snow) +
       aes(x = Week_snowmelt, y = Perc_snowcover) +
       geom_point()
```

(b)
```
ggplot(data = snow %>%
        group_by(Year) %>%
        summarise(
            avgPerc_snowcover = mean(Perc_snowcover),
            avgWeekSnowMelt = mean(Week_snowmelt))) +
        aes(x = avgWeekSnowMelt, y =
            ↳ avgPerc_snowcover) +
        geom_point()
```

(c)
```
ggplot(data = popsize) +
        aes(x = Herd, y = Pop_Size) +
        geom_boxplot()
```

(d)
```
# calculate summary stats
stats <- popsize %>%
        filter(Year >= 2008, Year <= 2014) %>%
        group_by(Year) %>%
        summarise(
            meanPopSize= mean(Pop_Size),
            SD = sd(Pop_Size),
            N = n())
# set up aesthetic mappings of error bars
limits <- aes(ymax = stats$meanPopSize + stats$SD,
            ymin = stats$meanPopSize - stats$SD)
# plot mean population size including SD
ggplot(data = stats) +
        aes(x = Year, y = meanPopSize) +
        geom_point() +
        geom_errorbar(limits, width = .3)
```

Intermezzo 9.6

(a)
```
ggplot(data = snow) +
        aes(x = Week_snowmelt) +
```

```
            geom_histogram() +
            facet_wrap(~Herd)
```

(b)
```
    ggplot(data = snow) +
            aes(x = Year, y = Herd, fill = Week_snowmelt)
                ↳  +
            geom_tile()
```

(c)
```
    ggplot(data = snow) +
            aes(x = Perc_snowcover, y = Week_snowmelt) +
            geom_point() +
            geom_smooth() +
            facet_wrap(~Herd)
```

Relational Databases

Intermezzo 10.1

(a) SELECT DISTINCT afr FROM lohr ORDER by CAST(afr AS
    INTEGER);
(b) SELECT pop, ro FROM lohr ORDER BY CAST(ro AS INTEGER)
    LIMIT 3;

Intermezzo 10.2

(a) SELECT COUNT(DISTINCT pop) FROM lohr;
(b) SELECT pop, AVG(CAST(afr AS INTEGER)) FROM lohr GROUP BY pop;
(c) SELECT pop, COUNT(ad) FROM lohr WHERE CAST(ad AS INTEGER) > 55
    GROUP BY pop;

Intermezzo 10.3

(a) SELECT Basin, AVG(Urban) FROM site GROUP BY Basin;
(b) SELECT DISTINCT Basin FROM site INNER JOIN trans
    ON site.SiteID = trans.SiteID WHERE anguilla = "extirpated"
    AND lucius = "extirpated" AND Per1 = "P3" AND Per2 = "P4";

Bibliography

● ● ● ● ● ● ● ● ● ● ●

Axelrod, R. (1980a). Effective choice in the prisoner's dilemma. *Journal of Conflict Resolution 24*(1), 3–25.

Axelrod, R. (1980b). More effective choice in the prisoner's dilemma. *Journal of Conflict Resolution 24*(3), 379–403.

Axelrod, R. and W. D. Hamilton (1981). The evolution of cooperation. *Science 211*(4489), 1390–1396.

Beitz, E. (2000). TeXshade: shading and labeling of multiple sequence alignments using LaTeX2e. *Bioinformatics 16*(2), 135–139.

Blischak, J. D., E. R. Davenport, and G. Wilson (2016). A quick introduction to version control with Git and GitHub. *PLoS Computational Biology 12*(1), e1004668.

Bolstad, G. H., J. A. Cassara, E. Márquez, T. F. Hansen, K. van der Linde, D. Houle, and C. Pélabon (2015). Complex constraints on allometry revealed by artificial selection on the wing of *Drosophila melanogaster. Proceedings of the National Academy of Sciences 112*(43), 13284–13289.

Buzzard, V., C. M. Hulshof, T. Birt, C. Violle, and B. J. Enquist (2016). Re-growing a tropical dry forest: Functional plant trait composition and community assembly during succession. *Functional Ecology 30*(6), 1006–1013.

Comte, L., B. Hugueny, and G. Grenouillet (2016). Climate interacts with anthropogenic drivers to determine extirpation dynamics. *Ecography 39*(10), 1008–1016.

Cumming, G., F. Fidler, and D. L. Vaux (2007). Error bars in experimental biology. *Journal of Cell Biology 177*(1), 7–11.

Dale, J., C. J. Dey, K. Delhey, B. Kempenaers, and M. Valcu (2015). The effects of life history and sexual selection on male and female plumage colouration. *Nature 527*(7578), 367–370.

Dalziel, B., O. Bjornstad, W. van Panhuis, D. Burke, C. Metcalf, and B. Grenfell (2016). Persistent chaos of measles epidemics in the prevaccination United States caused by a small change in seasonal transmission patterns. *PLoS Computational Biology 12*(2), e1004655.

Fauchald, P., T. Park, H. Tømmervik, R. Myneni, and V. H. Hausner (2017). Arctic greening from warming promotes declines in caribou populations. *Science Advances 3*(4), e1601365.

Fox, C. W., C. S. Burns, A. D. Muncy, and J. A. Meyer (2016). Gender differences in patterns of authorship do not affect peer review outcomes at an ecology journal. *Functional Ecology 30*(1), 126–139.

Gächter, S. and J. F. Schulz (2016). Intrinsic honesty and the prevalence of rule violations across societies. *Nature 531*(7595), 496–499.

Gesquiere, L. R., N. H. Learn, M. C. M. Simao, P. O. Onyango, S. C. Alberts, and J. Altmann (2011). Life at the top: Rank and stress in wild male baboons. *Science 333*(6040), 357–360.

Goldberg, E. E., J. R. Kohn, R. Lande, K. A. Robertson, S. A. Smith, and B. Igić (2010). Species selection maintains self-incompatibility. *Science 330*(6003), 493–495.

Jiang, Y., D. I. Bolnick, and M. Kirkpatrick (2013). Assortative mating in animals. *American Naturalist 181*(6), E125–E138.

Kacsoh, B. Z., Z. R. Lynch, N. T. Mortimer, and T. A. Schlenke (2013). Fruit flies medicate offspring after seeing parasites. *Science 339*(6122), 947–950.

Knauff, M. and J. Nejasmic (2014). An efficiency comparison of document preparation systems used in academic research and development. *PLoS ONE 9*(12), e115069.

Lahti, L., J. Salojärvi, A. Salonen, M. Scheffer, and W. M. de Vos (2014). Tipping elements in the human intestinal ecosystem. *Nature Communications 5*, 4344.

Letchford, A., H. S. Moat, and T. Preis (2015). The advantage of short paper titles. *Royal Society Open Science 2*(8), 150266.

Lohr, J. N. and C. R. Haag (2015). Genetic load, inbreeding depression, and hybrid vigor covary with population size: An empirical evaluation of theoretical predictions. *Evolution 69*(12), 3109–3122.

Malhotra, A., S. Creer, J. B. Harris, R. Stöcklin, P. Favreau, and R. S. Thorpe (2013). Predicting function from sequence in a large multifunctional toxin family. *Toxicon 72*, 113–125.

Marra, N. J. and J. A. DeWoody (2014). Transcriptomic characterization of the immunogenetic repertoires of heteromyid rodents. *BMC Genomics 15*(1), 929.

Martin, T. E. (2015). Age-related mortality explains life history strategies of tropical and temperate songbirds. *Science 349*(6251), 966–970.

Maynard Smith, J. (1982). *Evolution and the Theory of Games*. Cambridge University Press.

Mikaelyan, A., C. Dietrich, T. Köhler, M. Poulsen, D. Sillam-Dussès, and A. Brune (2015). Diet is the primary determinant of bacterial community structure in the guts of higher termites. *Molecular Ecology 24*(20), 5284–5295.

Miller, G. (2006). A scientist's nightmare: Software problem leads to five retractions. *Science 314*(5807), 1856–1857.

Pacifici, M., L. Santini, M. Di Marco, D. Baisero, L. Francucci, G. G. Marasini, P. Visconti, and C. Rondinini (2013). Generation length for mammals. *Nature Conservation 5*, 89–94.

Ram, K. (2013). Git can facilitate greater reproducibility and increased transparency in science. *Source Code for Biology and Medicine 8*(1), 7.

Saavedra, S. and D. B. Stouffer (2013). "Disentangling nestedness" disentangled. *Nature 500* (7463), E1–E2.

Schuster, P. and K. Sigmund (1983). Replicator dynamics. *Journal of Theoretical Biology 100*(3), 533–538.

Singh, N. D., D. R. Criscoe, S. Skolfield, K. P. Kohl, E. S. Keebaugh, and T. A. Schlenke (2015). Fruit flies diversify their offspring in response to parasite infection. *Science 349*(6249), 747–750.

Smith, F. A., S. K. Lyons, S. M. Ernest, K. E. Jones, D. M. Kaufman, T. Dayan, P. A. Marquet, J. H. Brown, and J. P. Haskell (2003). Body mass of late quaternary mammals: Ecological archives e084–094. *Ecology 84*(12), 3403–3403.

Taylor, P. D. and L. B. Jonker (1978). Evolutionary stable strategies and game dynamics. *Mathematical Biosciences 40*(1–2), 145–156.

Urban, M. C. (2015). Accelerating extinction risk from climate change. *Science 348*(6234), 571–573.

Wilkinson, L. (2006). *The Grammar of Graphics*. Springer Science & Business Media.

Wilson, G., D. Aruliah, C. Brown, N. Chue Hong, M. Davis, R. Guy, S. Haddock, K. Huff, I. Mitchell, M. Plumbley, B. Waugh, E. White, and P. Wilson (2014). Best practices for scientific computing. *PLoS Biology 12*(1), e1001745.

Żmihorski, M., T. Pärt, T. Gustafson, and Å. Berg (2016). Effects of water level and grassland management on alpha and beta diversity of wet grassland birds in restored wetlands. *Journal of Applied Ecology 53*(2), 587–595.

Index of Symbols

Index of Unix Commands

· ·

Index of Git Commands

Index of Python Functions, Methods, Properties, and Libraries

• •

Index of LATEX Commands and Libraries

· ·

Index of R Functions and Libraries

Index of SQLite Commands

General Index

• • • • • • • • • • • •

Italic pages refer to figures and tables